# IMF PROGRAMMES IN DEVELOPING COUNTRIES

# DEVELOPMENT POLICY STUDIES

Edited by John Farrington and Tony Killick
*for the Overseas Development Institute, London*

This series presents the results of ODI research on policy issues confronted by the governments of developing countries and their partners in aid and trade. It will be of interest to policy makers and practitioners in government and international organisations and to students and researchers in both North and South.

In this series:

IMF LENDING TO DEVELOPING COUNTRIES
*Graham Bird*

MANAGING WATER AS AN ECONOMIC RESOURCE
*James Winpenny*

# IMF PROGRAMMES IN DEVELOPING COUNTRIES

## Design and Impact

*Tony Killick*

London and New York

First published 1995
by Routledge
11 New Fetter Lane, London EC4P 4EE

Simultaneously published in the USA and Canada
by Routledge
29 West 35th Street, New York, NY 10001



Typeset in Garamond by
J&L Composition Ltd, Filey, North Yorkshire

Printed and bound in Great Britain by
Mackays of Chatham PLC, Chatham, Kent



*British Library Cataloguing in Publication Data*
A catalogue record for this book is available from the British Library

*Library of Contress Cataloguing in Publication Data*
A catalogue record for this book has been requested

ISBN 0–415–13040–9

# CONTENTS

# TABLES

# PREFACE

The Overseas Development Institute has a long-standing interest in studying – and improving – the ways in which the policies of international agencies impact upon welfare in developing countries, and this volume is a product of that interest. This, and its companion, follow from an earlier ODI project, which studied the impact of International Monetary Fund policies upon developing countries, utilising information primarily from the 1960s and 1970s. The purpose here is to trace the extent to which the Fund has changed over the last decade and a half and to re-examine the impact of the adjustment programmes which it sponsors on the basis of more up-to-date information.

I have considerably more than the usual help to acknowledge. Chapter 3 is the hard core of this volume and in preparing it I was extremely fortunate to have the help of outstanding assistants, Moazzam Malik and Marcus Manuel, who were my co-authors in an earlier presentation of some of the results in articles in *The World Economy*. I was no less fortunate in having the help of Ramani Gunatilaka in updating that work and subjecting it to further analysis. I should also offer special thanks to Graham Bird for his detailed comments on earlier versions of the text.

Much gratitude is also due to the UK Overseas Development Administration, which provided a grant to fund the research. The staff of the IMF were also generous, as they always are, in finding time to respond to my questions and providing documentation which would otherwise not have been easy to find. Margaret Cornell did her usual excellent job of editing the manuscript and Jane Horsfield did an immense amount of efficient work at all stages of the project, to bring it all into a condition fit to be shown to publishers.

My thanks to all the above, to the considerable number of friends who provided helpful comments on various parts of the text, and to my wife, Inge, for her loving willingness to tolerate and support a life driven by publishers' deadlines.

This volume assumes that the reader has a basic understanding of the purposes and modalities of the IMF and of the controversies surrounding it, but there are various easily accessible sources for such information.

The IMF itself publishes annually a useful factual description of its work in the form of a 'Supplement' to the issue of the *IMF Survey* brought out immediately prior to its annual meeting in late September or early October. Its *Annual Reports* also contain much valuable information. The IMF's Occasional Paper No. 55, *Theoretical Aspects of the Design of Fund-Supported Adjustment Programs* (1987), is also very useful on the principles of Fund conditionality. A brief description of the various 'facilities' (lines of credit) offered by the Fund is provided in the Appendix to Chapter 1.

As described in the main text, this book is a follow-up to a two-volume study published in 1984 but now unfortunately out of print. This was published by Gower Press, London as: Tony Killick (ed.), *The Quest for Economic Stabilisation: The IMF and the Third World* and *The IMF and Stabilisation: Developing Country Experiences.* These, particularly *The Quest*, provide useful background to the present study. So too does a more recent collection edited by Catherine Gwin and Richard Feinberg, *The International Monetary Fund in a Multipolar World: Pulling Together* (Washington, DC: Overseas Development Council, 1989). Jacques J. Polak's *Princeton Essay in International Finance* No. 184, 'The Changing Nature of IMF Conditionality' (1991) is also recommended.

This volume should be read alongside a companion volume by Graham Bird, *IMF Lending to Developing Countries: Issues and Evidence* (London: Routledge, 1995). The division of labour between the two volumes is that Bird concentrates more on the lending operations of the Fund, in the context of the world 'system' of international finance, while the present volume is focused on the design of the Fund's programmes – its 'conditionality'. Bird reviews the current state of the debate concerning the future role of the Fund, explores analytical issues concerning its financing and lending policies, and reviews the evidence relating to this, and presents a strategy for the future of its lending operations.

By comparison with *The Quest*, the conclusions of this volume are a good deal more sceptical about the possibilities of reforming policies by means of conditionality, and I have been stimulated to embark on a follow-on study of conditionality which is taking me well beyond the policies of the IMF. That will be the next publisher's deadline.

Tony Killick
November 1994

# ABBREVIATIONS

| | |
|---|---|
| BoP | Balance of Payments |
| CCFF | Compensatory and Contingency Financing Facility |
| EFF | Extended Fund Facility |
| ESAF | Enhanced Structural Adjustment Facility |
| NGO | Non-governmental Organisation |
| ODI | Overseas Development Institute |
| PFP | Policy Framework Paper |
| SAF | Structural Adjustment Facility |
| SAP | Structural Adjustment Programme |
| SDR | Special Drawing Right |

# 1

# STARTING POINTS

## THE FIRST BITE AT THE CHERRY

This book and its companion revisit a topic first examined (with Graham Bird, Jennifer Sharpley and Mary Sutton) in the early 1980s (see Preface). The back-drop to that study was one of acute global economic turmoil and fierce controversies about the policies of the IMF and their consequences. The second oil shock of 1979–80 (whose consequences for oil-importing developing countries were far more severe than the first, 1973, price explosion) and an associated recession among industrial countries had contributed to acute balance-of-payments difficulties in many developing countries. The world was marked by huge but rapidly changing payments disequilibria; large but unstable private banking flows; and uncertainties about the role of the IMF in an era of large imbalances and volatile exchange rates. The international financial system thrust much of the burden of adjustment to these forces upon deficit developing countries with no parallel leverage to exert upon those earning the counterpart surpluses, thus threatening the well-being of peoples already living below or close to the poverty line.

The IMF's attempts to help deficit countries were, at the same time, bitterly criticised as rigidly doctrinaire, forcing domestic adjustment in response to internationally generated difficulties, holding back development and imposing self-defeatingly harsh policy conditions. Macroeconomic management had been neglected in the literature on development, and it was still often seen as standing in opposition to long-term development.

The results of our study, published in 1984 as *The Quest for Economic Stabilisation: The IMF and the Third World* (hereafter *The*

*Quest*), questioned whether the fierceness of these controversies was justified by the facts. While agreeing with the IMF on the importance of macro management as an input into the development effort, it found that Fund programmes had limited impact in developing countries. They were subject to frequent breakdown, appeared to bring only moderate improvements to the balance of payments and were not systematically linked with trade liberalisation or with any strong deflationary effects. The IMF was shown as having difficulty in securing compliance with its policy conditions and as wielding only slight revealed influence on key policy variables. There was no more than a moderate connection between implementation and the achievement of programme objectives. Overall, to mix our Shakespeare, the sound and fury of the controversies about Fund programmes seemed much ado about nothing.

Since developing countries remained in urgent need of more effective balance-of-payments policies, the question arose whether there was not a better, lower-cost way of going about this task. The case for change was strong. The Fund's traditional emphasis on the control of aggregate demand was likely to be a high-cost solution to payments deficits originating in the gyrations of the world economy and/or from weaknesses in the productive structures of deficit countries. Conventional, short-term programmes were not well designed to cope with such 'structural' problems.

The changing geography of the Fund's lending added strength to the case for change, for an increasingly large share of all Fund programmes were in *low-income* developing countries, for whose circumstances the Fund's traditional short-term approach was least well suited. The case for change was further argued by recalling the Fund to its own Articles of Agreement, which specify the maintenance of high levels of employment, income and economic development as 'the primary objectives of economic policy' which it can assist through balance-of-payments support.

But while the case for change was strong, *The Quest* found only limited flexibility in the Fund. The IMF had sought to adapt to the problems caused by the oil shocks and world recession by the creation of new facilities, some movement towards longer-term lending and some easing in its conditionality, but it had done little to change the nature of its programmes. In any case, these experiments had been put sharply into reverse at the beginning of the 1980s, 'so that by 1982 its conditionality seemed rather similar to that of the 1960s' (*The Quest*: 222). There had been some attempt to adapt but we described this as a 'constrained flexibility' and drew attention to the strength of internal and external resistances to change.

We saw the *costs* of adjustment as the essential problem and urged the Fund towards a cost-minimising 'real economy' approach. This would place greater weight on supply-side measures intended to stimulate output and productivity, and go beyond conventional macro aggregates to include measures of a more microeconomic kind in order to resolve bottlenecks and rigidities within the productive system. It would, of course, entail longer-term programmes and more supporting finance.

## THE CASE FOR A SECOND BITE

That was how we saw things, writing in the early 1980s. Ten years later there seemed good reason to take a second look. For one thing, the decade had seen major developments in the international monetary system. Instability had increased in the exchange rates of the major currencies, with attendant costs to world trade. There had been important developments in currency arrangements, notably in the European Monetary System. In 1982 the 'debt crisis' had struck, forcing heavily indebted countries into years of hardship and inducing large shifts in the geographical pattern of international bank lending, including a steep fall in commercial lending to developing countries. The USA had emerged as the world's largest debtor country, with considerable world-wide consequences. International liquidity creation had become effectively privatised, causing the near-demise of the SDR. A number of these developments called into question the ability of the IMF to carry out its Articles of Agreement and, indeed, the continuing relevance of that mandate. There were thus doubts about its future.

The period had also witnessed a remarkable turn-around in the Fund's financial relationships with developing countries, with a swing from being a large net provider of assistance to deficit countries to being a net recipient of return flows. There was the associated emergence of a serious problem of arrears to the Fund by member countries, leading to some of them being declared ineligible for further Fund assistance. Most recently a new group of borrowing countries has emerged – the 'countries in transition' of Eastern Europe and the former Soviet Union. In addition to a

*Table 1.1* IMF–Member agreements in effect at any time during the period 1988/89–1992/93 – number of agreements and amounts allocated, developing countries and countries in transition

| *Number of Agreements* | *1988/89* | *1989/90* | *1990/91* | *1991/92* | *1992/93* | *Total* |
|---|---|---|---|---|---|---|
| Developing countries | | | | | | |
| Stand-bys | 10 | 13 | 8 | 21 | 3 | 55 |
| EFFs | 1 | 3 | 0 | 1 | 4 | 9 |
| SAFs excluding ESAF conversions | 4 | 3 | 1 | 1 | 1 | 10 |
| ESAFs | 5 | 4 | 3 | 5 | 8 | 25 |
| Total | 20 | 23 | 12 | 28 | 16 | 99 |
| Countries in transition | | | | | | |
| Stand-bys | 2 | 3 | 11 | 3 | 6 | 25 |
| EFFs | 0 | 0 | 2 | 0 | 0 | 2 |
| SAFs excluding ESAF conversions | 0 | 0 | 0 | 0 | 0 | 0 |
| ESAFs | 0 | 0 | 0 | 0 | 0 | 0 |
| Total | 2 | 3 | 13 | 3 | 6 | 27 |
| *Percentage of total number of programmes in developing and transition countries* | | | | | | |
| Developing countries | | | | | | |
| Stand-bys | 83 | 81 | 42 | 88 | 33 | 69 |
| EFFs | 100 | 100 | 0 | 100 | 100 | 82 |
| SAFs excluding ESAF conversions | 100 | 100 | 100 | 100 | 100 | 100 |
| ESAFs | 100 | 100 | 100 | 100 | 100 | 100 |
| Total | 9 | 88 | 48 | 90 | 73 | 79 |
| Countries in transition | | | | | | |
| Stand-bys | 17 | 19 | 58 | 13 | 67 | 31 |
| EFFs | 0 | 0 | 100 | 0 | 0 | 18 |
| SAFs excluding ESAF conversions | 0 | 0 | 0 | 0 | 0 | 0 |
| ESAFs | 0 | 0 | 0 | 0 | 0 | 0 |
| Total | 9 | 12 | 52 | 10 | 27 | 21 |

| | | | | | | |
|---|---|---|---|---|---|---|
| *Amounts allocated* (SDRm) | | | | | | |
| Developing countries | | | | | | |
| Stand-bys | 2,377 | 2,085 | 591 | 5,057 | 125 | 10,235 |
| EFFs | 207 | 7,627 | 0 | 2,149 | 1,586 | 11,569 |
| SAFs excluding ESAF conversions | 441 | 45 | 31 | 3 | 49 | 570 |
| ESAFs | 955 | 415 | 527 | 637 | 478 | 3,012 |
| Total | 3,980 | 10,172 | 1,149 | 7,846 | 2,238 | 25,386 |
| Countries in transition | | | | | | |
| Stand-bys | 562 | 1,164 | 1,279 | 441 | 1,342 | 4,789 |
| EFFs | 0 | 0 | 2,338 | 0 | 0 | 2,338 |
| SAFs excluding ESAF conversions | 0 | 0 | 0 | 0 | 0 | 0 |
| ESAFs | 0 | 0 | 0 | 0 | 0 | 0 |
| Total | 562 | 1,164 | 3,617 | 441 | 1,342 | 7,127 |
| *Percentage of total amount allocated in developing and transition countries* | | | | | | |
| Developing countries | | | | | | |
| Stand-bys | 81 | 64 | 32 | 92 | 9 | 68 |
| EFFs | 100 | 100 | 0 | 100 | 100 | 83 |
| SAFs excluding ESAF conversions | 100 | 100 | 100 | 100 | 100 | 100 |
| ESAFs | 100 | 100 | 100 | 100 | 100 | 100 |
| Total | 88 | 90 | 24 | 95 | 63 | 78 |
| Countries in transition | | | | | | |
| Stand-bys | 19 | 36 | 68 | 8 | 91 | 32 |
| EFFs | 0 | 0 | 100 | 0 | 0 | 17 |
| SAFs excluding ESAF conversions | 0 | 0 | 0 | 0 | 0 | 0 |
| ESAFs | 0 | 0 | 0 | 0 | 0 | 0 |
| Total | 12 | 10 | 76 | 5 | 37 | 22 |

*Source*: Calculated from IMF, *Annual Reports* (various issues).

*Note*: Classification of countries in transition according to IMF *World Economic Outlook*, May 1993, p. 123.

'Systemic Transformation Facility' hurriedly created for them in 1993, this group has also received a substantial number of conventional stand-by credits. Nevertheless, developing countries remain the Fund's chief customers, accounting in recent years for four-fifths of its lending, both in numbers of credits and in value terms (see Table 1.1).[1]

This volume concentrates on the design and conditionality of Fund programmes. In this area too a great deal has happened since the beginning of the 1980s, raising the question whether the Fund has been able to respond adequately both to the felt deficiencies revealed in *The Quest* and to the changing circumstances of the time.

The emergence of the debt problem in the early 1980s was a traumatic development and exerted a decisive influence on the policies of the IMF. In the early years the debt situation posed a serious threat to the whole structure of international banking. It thus impinged forcefully on the national interests of some of the Fund's most powerful shareholders, most notably the United States, and eventually made them open to policy shifts which previously they had resisted.

The effects of the large debt-servicing burdens of the heavily indebted countries, and their greatly reduced access to world capital markets, shifted the adjustment–financing balance sharply towards adjustment, and in most of these countries it was a task of the Fund to ensure that sufficiently 'rigorous' policy responses were put in place. Initially, when debtor countries were viewed as essentially facing a liquidity problem, the rapid spread of Fund programmes among them was not seen as posing any novel issues for the design of conditionality; it was more a matter of administering the familiar medicine in larger doses and with a thinner sugar coating.

However, it gradually came to be recognised that the immediate post-1982 response was unviable. Indebted countries were pushed into prolonged recessions, with most of the benefits of often fiercely deflationary demand-management policies seen as accruing to foreign commercial bankers and other creditors, and with the Fund increasingly perceived as playing the role of debt collector. The political unsustainability – and perhaps the economic undesirability – of this situation gradually dawned, not least on the then US Treasury Secretary, James Baker. The 1985 'Baker Plan', based on the premise that 'Sustainable growth with adjustment must . . . be the central objective of our debt strategy', brought a marked shift in policy (even though the 'plan' itself bore few fruit). Not the least surprising aspect of this shift was that it brought the US Administration into the camp of those vocally critical of anti-growth Fund programmes. 'Adjustment with growth' became the buzz-word of the day, with the Fund staff under pressure to prepare 'growth-oriented' programmes.

The Baker Plan was chiefly addressed to the positions of countries, mostly Latin American, which had borrowed heavily from commercial banks. Concurrently, there was also growing concern about the position of heavily indebted low-income 'official borrowers', chiefly in sub-Saharan Africa, whose debt situations continued to worsen. The manifest unviability of creditor policies towards this group of countries combined with political and humanitarian considerations to motivate new initiatives which would alleviate their plight. (For a good account of the connections between the debt difficulties of African countries and changes in IMF policies, see Martin, 1991: Chapter 7.)

Other forces were also working on the Fund. One was the rapid growth during the first half of the 1980s of 'structural adjustment' lending by the World Bank. This initiative carried the Bank into IMF territory, by giving it a 'seat at the table' in macroeconomic policy formation in debtor countries and by offering a form of balance-of-payments support that appeared to respond more directly to the structural nature of the payments weaknesses of many developing countries. This change called for a Fund response, to ensure that it was not overshadowed by the Bank and that it retained its lead role in macroeconomic policy. Because of the growing overlap between the two institutions, it also had to work out modalities of co-operation with the Bank, to reconcile its own philosophy of balance-of-payments management with the latter's more structurally oriented approach.

Yet another element to which the IMF had to respond was the emergence for the first time of an important group of countries which had fallen into arrears in servicing its past credits. In the earlier study we were able to report (*The Quest*, pp. 184–5) that 'the Fund has never suffered a formal default of more than a technical or transitional nature, although some accounting acrobatics and some rolling-over have been necessary to avoid a default'. This situation changed dramatically during the 1980s. The Fund's *Annual Reports* show three countries to have fallen into arrears to the extent of SDR 0.3bn by April 1984, figures which had grown to eleven countries and SDR 3.4bn by April 1990 – an amount which can be compared with total outstanding Fund credit at the latter date of SDR 24.4bn.[2] A deterioration of this magnitude put considerable pressure on the Fund, raising questions about the soundness of its lending decisions and for the first time throwing doubts on the quality of its balance sheet. As we shall see later, this exerted a substantial influence on the institution's policies in the second half of the decade.

These symptoms of the straitened finances of many of its developing country members thus called into question the appropriateness of the Fund's traditional policy stances. The well-researched recommendations of a 1987 G-24 report added to the impetus for policy reform (see Chapter 2), as did the growing volume of academic evaluations of the Fund's performance and policies.[3] In contrast to the neglect of the 1960s and 1970s, there was a burst of publications during the 1980s dealing with macroeconomic management and adjustment in developing countries, and with the IMF itself. At the same time, resistance to reform within the IMF's staff and Executive Board remained strong. It was therefore important to establish the extent to which it had actually responded to the pressures for change. There thus appeared a need for an examination of the arguments and evidence which was synoptic and (so far as humanly possible) objective. The Fund's 50th Anniversary in 1994 provided a further occasion for taking stock of present knowledge to see how the Fund's role might develop into the twenty-first century.

## THE STRUCTURE OF THE BOOK

A good many of the systemic issues raised above are dealt with in Graham Bird's companion volume. This book concentrates more narrowly on the content and consequences of Fund programmes. Chapter 2 takes up the question just posed, of the extent to which programme design has been adapted to the conditions summarised above. A lengthy Chapter 3 gathers together the considerable body of evidence which has by now accumulated on the effects of Fund programmes, as well as reporting the results of new research, and offers a synoptic view of the conclusions which emerge from this pooling of information. Chapter 4 is more theoretical, setting out the basic model underlying Fund conditionality and subjecting this to critical evaluation. Chapter 5 examines the main conclusions emerging from the study and points to future directions for the Fund.

## APPENDIX: THE FACILITIES OF THE IMF[4]

### Purchases and repurchases

The normal procedure is for a member country to make a purchase, or drawing, from the IMF by exchanging its own currency for an equivalent amount of other members' currencies or Special Drawing Rights (SDRs) held by the Fund. Over a prescribed period, it is then required to repurchase – buy back – its own currency with other members' currencies or SDRs. The IMF levies charges on purchases, and the net effect of a member's purchase and subsequent repurchase is very similar to that of its receiving a loan at interest and subsequently repaying it.

### First tranche credits

To qualify for a first tranche credit a member government must demonstrate reasonable efforts to overcome balance-of-payments difficulties during the programme period. Purchases are not phased and are not subject to rigorous conditionality, including performance criteria. Repayments (repurchases) are made in $3\frac{1}{4}$–5 years.

### Stand-by credits

Here, a member must have a substantial and viable programme to overcome its balance-of-payments difficulties. Assistance is normally provided in the form of stand-by credits that provide purchases in instalments linked to the observance of previously specified performance criteria. Repurchases are made in $3\frac{1}{4}$–5 years.

### Extended fund facility

EFF credits are for a medium-term programme aimed at overcoming structural balance-of-payments maladjustments. A programme generally lasts for three years, although it may be lengthened to four years. The programme initially identifies policies and measures for the first 12-month period in detail. Resources are provided in the form of extended arrangements that include performance criteria and drawings in instalments. Repurchases are made in 4½–10 years.

### Structural adjustment facility

SAF resources are provided as loans on concessional terms to low-income developing countries facing protracted balance-of-payments problems, in support of medium-term macroeconomic and structural adjustment programmes. A medium-term policy framework for a three-year period is prepared jointly by the member government, the Fund and the World Bank, and is set out in a policy framework paper (PFP). Detailed annual programmes are formulated prior to the disbursement of loans under annual arrangements that include quarterly benchmarks used to assess performance, but conditionality is relatively slight. Interest is 0.5 per cent annually, and repayments are made in 5½–10 years.

### Enhanced structural adjustment facility

The objectives, eligibility, terms and basic programme features of ESAF parallel those of the SAF; differences relate to provisions for access, monitoring, and funding, and conditionality is far more rigorous. A PFP and a detailed annual programme are prepared each year. Arrangements under the ESAF include quarterly benchmarks, semi-annual performance criteria, and a mid-year review. Adjustment measures are expected to be particularly strong, aiming to foster growth and to achieve a substantial strengthening of the member's balance-of-payments position. Loans are disbursed semi-annually, and repayments are made in 5½–10 years.

### Compensatory and contingency financing facility

The compensatory elements of the CCFF provide resources to a member to help compensate for a shortfall in export earnings or an excess in cereal import costs that is temporary and is owing to factors largely beyond the member's control. The contingency element helps members with IMF-supported adjustment programmes to maintain the momentum of adjustment efforts in the face of unanticipated, adverse external shocks. Repurchases are made in $3\frac{1}{4}$–5 years.

### Systemic transformation facility

Created in 1993 specially to assist the countries of the former Soviet Union and other East European countries, this temporary facility provides resources to member countries facing balance-of-payments difficulties arising from severe disruptions of their traditional trade and payments arrangements owing to a shift from significant reliance on trading at non-market prices to multilateral, market-based trade. The facility is designed to provide assistance to countries that co-operate fully with the IMF towards finding appropriate solutions to their balance-of-payments problems but who may not yet be able to formulate a programme under the IMF's regular facilities, i.e. conditionality is relatively slight. Half of the total financing provided is disbursed at the outset, with the remainder normally disbursed in about 6 months, but not later than 12 months, after the first purchase, provided the applicable conditions are met. Repurchases are made in $4\frac{1}{2}$–10 years.

# 2

# CONTINUITY AND CHANGE IN IMF PROGRAMME DESIGN, 1982–92

To the case for change summarised in Chapter 1 should be added a considerable body of intellectual criticism of the design and appropriateness of IMF programmes, and it is with these criticisms that we begin.

## RESPONDING TO CRITICISMS

### A statement of the controversy

Much of the case of those who complain about the inappropriateness of Fund conditionality to developing country circumstances is encapsulated in a report by the Group of Twenty-Four (G-24, 1987: 9):

> The experience of developing countries that have undertaken Fund supported adjustment programs has not generally been satisfactory. The Fund approach to adjustment has had severe economic costs for many of these countries in terms of declines in the levels of output and growth rates, reductions in employment and adverse effects on income distribution. A typical Fund program prescribes measures that require excessive compression of domestic demand, cuts in real wages, and reductions in government expenditures; these are frequently accompanied by sharp exchange rate depreciation and import liberalization measures, without due regard to their potentially disruptive effects on the domestic economy.

Underlying the complaint, not confined to the G-24, about excessive reliance on demand compression, is the view that the

IMF is too much of a monetary organisation, dogmatically applying a monetarist approach to balance-of-payments policy in standardised ways which pay insufficient heed to individual country circumstances, and that it is inflexible in its dealings with debtor governments. Fund programmes are thus seen as too short-term, and neglecting the more structural, supply-side causes of payments difficulties. Indeed, the cuts in public investment with which its programmes are associated are seen as positively detrimental to the process of structural adaptation, both directly and indirectly through their discouraging effect on private investment (the 'crowding-in' argument).[1]

The Fund has also been criticised for insensitivity. As Helleiner (1986: 8) reported the views of others: 'the Fund staff is inadequately informed or insensitive with respect to local conditions and objectives, patronising in their relationships with local professionals, and rigid or powerless or both in their negotiations'. Its critics emphasis the deflationary and poverty-increasing nature of its policy preferences and it is complained that the design of Fund programmes is ill-suited for the specific conditions of low-income countries. Loxley (1986: 121–2), for example, argues that the lesser responsiveness of their economies means that devaluations will bring smaller, and slower, payments benefits; and Chapter 4 below suggests that the assumptions about the effectiveness of monetary policy implicit in the Fund financial programming model are questionable in the conditions of low-income countries.

The extent to which the Fund has been able to adapt its position within the wider international monetary system is examined by Graham Bird in the companion volume. Our task here is the narrower one of examining the extent to which the Fund has responded since the early 1980s to the criticisms just described, and to the pressures for change outlined in Chapter 1 as they relate to the appropriateness for developing country circumstances of its facilities and of the design of its adjustment programmes.

## The changing pattern of facilities

The Fund's most notable response was the creation of new facilities (see Chapter 1, Appendix), designed specifically to provide medium-term assistance to low-income debtor countries facing protracted payments difficulties, thereby accepting the principle that such countries required special provisions. The first of these was the Structural Adjustment Facility (SAF), set up in 1986 and financed by reflows of Trust Fund money. SAF credits are limited to a maximum of only 70 per cent of a country's quota but are available at a heavily subsidised interest rate and repayable over up to 10 years. They are in support of medium-term (i.e. three-year) programmes and the conditionality attached to them is not very demanding. Most SAF loans have been to low-income African member states.

The SAF was followed by the creation in 1987 of the Enhanced Structural Adjustment Facility (ESAF), which, although also restricted to low-income countries, has had greater influence. Early in 1994 this facility was renewed on an enlarged scale. ESAF credits are in support of what the Fund calls 'especially vigorous' medium-term adjustment programmes, available on the same soft financial terms as SAF but generally much larger in size.

The effect of the opening of the SAF and ESAF windows on the overall pattern of Fund lending is shown in Tables 2.1 and 2.2. Table 2.2 shows that SAF and ESAF programmes made up over half the total number of credits outstanding in March 1994. However, Table 2.1 helps to keep the significance of these new facilities in perspective, for it shows that at all times the aggregate value of stand-by lending has greatly exceeded SAF and ESAF lending, and that EFF credits became large at the end of the period. In value terms SAF and ESAF credits accounted for only a third of the total, because they were relatively small, being for poor and usually small countries. However, ESAF credits comprised a high proportion of lending to low-income countries, and most lending in sub-Saharan Africa.

By common consent, performance and monitoring provisions are much more stringent with ESAF credits, which are supposed to be linked to 'especially vigorous' policy reforms. In the diplomatic language of the IMF, 'Although disbursements are not directly related to the observance of benchmarks, deviations would indicate the need for policy adjustments under the subsequent annual program' (*IMF Survey* Supplement, 1993: 2). Continuing access to ESAF, but not SAF, money is conditional on the observance of performance criteria.

Another significant development was the conversion through a series of decisions during the 1980s of the long-standing Compensatory Financing Facility from a more or less automatic fund for largely non-conditional assistance to countries experiencing

*Table 2.1* Composition of IMF lending by facility, 1980/81 to 1992/93[a]

| | Commitments during the year | | | | | | | | | Total outstanding commitments at year-end | | | | | | | | |
|---|---|---|---|---|---|---|---|---|---|---|---|---|---|---|---|---|---|---|
| | *Stand-bys* | | *EFF* | | *SAF* | | *ESAF* | | *Total* | *Stand-bys* | | *EFF* | | *SAF* | | *ESAF* | | *Total* |
| | *SDRm* | % | *SDRm* | % | *SDRm* | % | *SDRm* | % | *SDRm* | *SDRm* | % | *SDRm* | % | *SDRm* | % | *SDRm* | % | *SDRm* |
| 1980/81 | 5,197 | 50 | 5,221 | 50 | – | – | – | – | 10,419 | 5,331 | 49 | 5,464 | 51 | – | – | – | – | 10,795 |
| 1981/82 | 3,106 | 28 | 7,908 | 72 | – | – | – | – | 11,014 | 6,296 | 39 | 9,910 | 61 | – | – | – | – | 16,206 |
| 1982/83 | 5,450 | 39 | 8,671 | 61 | – | – | – | – | 14,121 | 9,464 | 38 | 15,561 | 62 | – | – | – | – | 25,025 |
| 1983/84 | 4,287 | 98 | 95 | 2 | – | – | – | – | 4,382 | 5,448 | 29 | 13,121 | 71 | – | – | – | – | 18,569 |
| 1984/85 | 3,218 | 100 | – | – | – | – | – | – | 3,218 | 3,925 | 34 | 7,750 | 66 | – | – | – | – | 11,675 |
| 1985/86 | 2,123 | 72 | 825 | 28 | – | – | – | – | 2,948 | 4,076 | 83 | 831 | 17 | – | – | – | – | 4,907 |
| 1986/87 | 4,118 | 89 | – | – | 488 | 11 | – | – | 4,605 | 4,313 | 80 | 750 | 14 | 327 | 6 | – | – | 5,391 |
| 1987/88 | 1,701 | 58 | 245 | 8 | 1,009 | 34 | – | – | 2,956 | 2,187 | 48 | 995 | 22 | 1,357 | 30 | – | – | 4,540 |
| 1988/89 | 2,956 | 65 | 207 | 5 | 441 | 10 | 955 | 21 | 4,560 | 3,054 | 46 | 1,032 | 16 | 1,566 | 24 | 955 | 14 | 6,608 |
| 1989/90 | 3,249 | 29 | 7,627 | 67 | 45 | – | 415 | 4 | 11,337 | 3,597 | 26 | 7,834 | 56 | 1,110 | 8 | 1,370 | 10 | 13,911 |
| 1990/91 | 2,786 | 50 | 2,338 | 42 | 53 | 1 | 426 | 8 | 5,603 | 2,703 | 18 | 9,597 | 65 | 539 | 4 | 1,813 | 12 | 14,652 |
| 1991/92 | 5,587 | 64 | 2,493 | 29 | 3 | – | 637 | 7 | 8,720 | 4,833 | 25 | 12,159 | 63 | 101 | 1 | 2,111 | 11 | 19,203 |
| 1992/93 | 1,971 | 53 | 1,242 | 33 | 49 | 1 | 478 | 13 | 3,740 | 4,490 | 29 | 8,569 | 56 | 83 | 1 | 2,137 | 14 | 15,279 |
| 1993/94 | 1,381 | 41 | 779 | 23 | 27 | 1 | 1,192 | 35 | 3,379 | 1,131 | 13 | 4,504 | 53 | 80 | 1 | 2,713 | 32 | 8,428 |

*Source*: IMF *Annual Reports*.

*Note*: [a] Years ending 30 April.

*Table 2.2* IMF Commitments as at March 1994

| | *Stand-bys* | | | *EFFs* | | | *SAFs* | | | *ESAFs* | | | *Total* | | |
|---|---|---|---|---|---|---|---|---|---|---|---|---|---|---|---|
| | *No.* | *SDRm* | *%* | *No.* | *SDRm* | *%* | *No.* | *SDRm* | *%* | *No.* | *SDRm* | *%* | *No.* | *SDRm* | *%* |
| Low-income countries | 4 | 197 | 2 | 3 | 894 | 10 | 3 | 107 | 1 | 18 | 2,133 | 24 | 28 | 3,330 | 37 |
| *of which* sub-Saharan Africa | 3 | 52 | 1 | 1 | 115 | 1 | 3 | 107 | 1 | 13 | 1,081 | 12 | 20 | 1,354 | 15 |
| Other developing countries | 5 | 289 | 3 | 3 | 3,610 | 41 | 1 | 3 | – | 2 | 497 | 6 | 11 | 4,399 | 50 |
| Total developing countries | 9 | 486 | 5 | 6 | 4,504 | 51 | 4 | 110 | 1 | 20 | 2,630 | 30 | 39 | 7,729 | 87 |
| Countries in transition | 8 | 1,079 | 12 | – | – | – | – | – | – | 2 | 83 | 1 | 10 | 1,162 | 13 |
| Total | 17 | 1,565 | 17 | 6 | 4,504 | 51 | 4 | 110 | 1 | 22 | 2,713 | 31 | 49 | 8,891 | 100 |

*Source*: IMF *Memorandum*, 2 May 1994.

temporary export shortfalls for reasons beyond their control into a modest source of supplementary finance mainly confined to member countries which have already agreed a stand-by or other high-conditionality programme. In effect, this has largely become another high-conditionality facility. As access to this facility was tightened, however, its coverage was widened, first to include temporary increases in cereal import requirements, usually arising because of climatically induced harvest failures, and then to incorporate a contingency mechanism offering additional finance to countries whose stand-by or other programmes ran into difficulties because of unforeseen deteriorations in export or import prices or in international interest rates. This facility is now called the Compensatory and Contingency Financing Facility (CCFF).

Implicit in these alterations to the CFF was a change in the balance between the intensity of the adjustment effort and the provision of finance. This balance shifted sharply during the 1980s, chiefly as a result of the emergence of the debt crisis. After 1982 there was an abrupt cessation of net commercial bank lending to most developing countries and the banks began endeavours to reduce their exposure. There being no comparable increase in flows of public monies, indebted countries judged that they had little choice but to embark upon often severely deflationary fiscal and monetary policies, a feature which had a particularly acute effect on the heavily indebted Latin American countries.

The IMF responded to this situation by modifying its procedures for stand-bys so that, for heavily indebted countries, it could delay programme approval or activation until the commercial banks had entered into specific commitments on debt relief. This increased the pressures on the banks, whose outstanding loans were unlikely to be serviced by the debtor country until the programme became operative. Such provisions were quite common for a few years from 1983, although they were unable to do more than slow down the rate at which the banks secured a net return flow from the debtors. Early in the 1990s, the Fund went further and partially relaxed its previous insistence that countries must avoid arrears with the banks as part of programme conditionality, in a move intended to reduce the extent to which it appeared to be acting as debt collector for the banks.

After 1982 the Fund also greatly increased its own lending to indebted countries and the rules governing the maximum size of its credits were relaxed, so that in exceptional cases there was no formal upper limit at all. For all that, however, the average size of Fund credits declined from an average of 158 per cent of quota in the period 1979–83 to 64 per cent during 1984–90 (calculated from Kafka, 1991: Table II). Moreover, its interest charges on ordinary credits (i.e. other than for the subsidised SAF/ESAF facilities) rose sharply with world capital market rates, and the grant element in its rates was almost eliminated (see Bird, 1995: Figure 3.3).

Lastly, as an example of Fund adaptation (albeit one forced upon it at US and other industrial countries' insistence) mention should be made of the setting-up in 1993 of the Systemic Transformation Facility, described in the Appendix to Chapter 1. However, since this was created specifically to channel assistance to the former Soviet Union and other ex-communist East European countries, rather than for developing countries *per se*, it lies outside the focus of this volume.

## CHANGES IN POLICY APPROACHES

Given this evolution in the facilities and lending patterns of the Fund, the issue that concerns us next is the extent to which the institution responded to the criticisms outlined earlier concerning the design of its programmes. To what extent has its conditionality also evolved? We begin our examination of this question by looking for changes in stated programme objectives.

### Programme objectives

The Fund has long described the objective of its programmes as to restore *viability* to the borrowing country's balance of payments. An authoritative early 1980s gloss on this was provided by Guitian (1981: 24):

> The concept of a viable balance of payments typically means, especially for many developing countries, a current account deficit that can be financed, on a sustainable basis, by net capital inflows on terms that are compatible with the development and growth prospects of the country.

(Note that the only reference to growth and development pertained to the terms on which external borrowing was undertaken.)

Similar language is still employed but with subtle variations. For example, a 1989 internal staff review states that the main programme objective is to achieve 'a sustainable balance of payments over the medium term in the context of an open trade and payments system'. As we shall see shortly, significance attaches to the references both to the medium term and to payments liberalisation. No less significant was the same document's reference to domestic social and political objectives, which went beyond traditional Fund ritual to accept that there were circumstances in which such objectives could act as a 'basic constraint' on programme design.

In the case of ESAF programmes the objective has been formally redefined so as to elevate economic growth to a primary objective. This upgrading of the growth objective has been one of the more notable changes of recent years, although we will later give reasons for doubting how much substance lies behind it. The Fund traditionally took the view that its primary task was to strengthen a country's payments position and that, by doing so, it was 'laying the foundations' upon which improved growth could be achieved in the future (*The Quest*, p. 187). The then Managing Director, Jacques de Larosière, stated bluntly in 1986 that 'economies beset with widespread price distortions, misaligned exchange rates, rampant inflation, pervasive trade restrictions, large budget deficits, heavy external debt, and wholesale capital flight simply cannot and do not grow rapidly for any sustained period of time'.[2] As late as 1987 the Group of Twenty-Four report was much concerned to persuade the IMF to upgrade the growth objective, in effect arguing that minimum growth targets should constrain the programme designs. By 1990, however, Managing Director Camdessus was using very different language from that of his predecessor: 'Our prime objective is growth. In my view, there is no longer any ambiguity about this. It is towards growth that our programs and their conditionality are aimed.'[3]

The Fund has also responded to the urgings of UNICEF and others in the direction of giving its programmes more of a 'human face'. The traditional Fund position was that it was a matter for the government to concern itself with the distributional impact of stabilisation programmes and that it would be inappropriate for the Fund to get involved in such fields. It was as good as its word. In a review of 30 stand-by programmes during the 1960s and 1970s, our earlier work found only one which contained provisions to protect the poor against possible adverse consequences (*The Quest*, Table B, p. 225).

This position has gradually changed. Fund missions now commonly discuss distributional aspects with governments when preparing programmes, and there is a requirement that the Policy Framework Papers prepared in connection with SAF/ESAF programmes should 'identify measures that can help cushion the possible adverse effects of certain policies on vulnerable groups . . . in ways consistent with the program's macroeconomic framework' (IMF, *Annual Report 1991*: 51–2). An increasing number of programmes now contain safety-net provisions for such groups, although the chief examples are in the former communist states of Eastern Europe, not in developing countries.

A further facet of the Fund's greater sensitivity to concerns going beyond the balance of payments is the increasing interest it is taking in the level of military spending by governments coming to it for assistance. This was another area previously regarded as off-limits in programme negotiations, and even now it remains outside the Fund's formal conditionality. However, according to Polak (1991: 29), in at least one country it sought and obtained assurances about plans for military spending and had exerted pressure to reduce such expenditures in a few other cases. We shall see later that the institution has, in general, been taking more interest in the detailed composition of government expenditures, but that it should be doing so in such a sensitive area is significant.

In another area the Fund has been more resistant to change, namely the environmental impact of its programmes. Under pressure from the environmentalist lobby, the US Administration and some other shareholders have been pushing the Fund to address this issue. But when the Managing Director brought proposals before the Board he was rebuffed by representatives of most major European shareholders and others, who saw a danger that macroeconomic conditionality could be diluted in exchange for some concessions on environmental matters that were largely irrelevant to the Fund's remit. The Board thus contented itself with the conclusion that the staff 'should be mindful of the interplay between economic policies, economic activity, and environmental change', and that it should avoid policies that could have undesirable environmental consequences in ways 'consistent with the Fund's mandate, size, and structure' (IMF, *Annual Report, 1991*: 54). In short, it stonewalled.

One other, unstated, aspect of balance-of-payments 'viability' which has also assumed greater importance in recent years is a requirement that the borrowing country should be able to service the IMF's own credits. We saw earlier how arrears in interest and amortisation payments to the Fund had emerged as a serious concern during the 1980s. According to one senior staff member, this led the Fund to come close to defining viability in terms of the likelihood that the country would be able to repay Fund credits as they fell due. Operational guidelines issued to the staff in 1988 emphasised the aim of limiting current account deficits to levels that could be financed by capital inflows which would not jeopardise the country's debt-servicing position, and the desirability of providing for an accumulation of international reserves to protect against adverse contingencies and to 'ensure the capacity to repay the Fund'.

Finally, we can revert to the reference in the 1989 internal review to the goal of achieving balance-of-payments viability 'in the medium term'. This too signals a shift of emphasis. The Articles of Agreement empower the IMF to make its resources 'temporarily available' to member countries seeking to strengthen their balance of payments, and part of its traditional theology has been that it was a credit co-operative whose resources were used on a 'revolving fund' basis. In the past this made the Fund reluctant to accept overtly a situation in which a country was making repeated use of its facilities, and rather insistent that programmes should aim for quick results. In 1981 the Managing Director stated that, wherever possible, balance-of-payments viability should be restored during the programme period (typically 12 to 18 months in the case of stand-bys); more often the restoration of viability was linked to the stand-by repayment period of three to five years.

In recent years, however, the Fund has come to recognise the tensions between its insistence on short-term credits and programmes and the provision of effective support to countries whose problems are deeply rooted in structural weaknesses. It now acknowledges that the restoration of balance-of-payments viability will often take a long time, especially in low-income countries. At least in the case of ESAF programmes, it now only requires the objective that 'substantial progress' should be made within a three-year period. Indeed, so far has this lengthening of time horizons gone that one authoritative commentator has concluded that the idea of the Fund as a credit union has been 'wholly overtaken' (Polak, 1991: 2).

The Fund has similarly moved further to a *de facto* acceptance of the repeated use of its facilities by developing country members, further diluting the principle that it can only lend on a 'temporary' basis. Although it formally does not encourage more than three successive programmes, the Board is willing to go along with prolonged use so long as this is carefully justified in the papers presented to it for decision. In fact, various countries have made repeated use of Fund credits. Chapter 3 reports the results of studies of a sample of 17 countries that borrowed from the Fund in the period 1979–89; of these no less than eight had a minimum of six programmes or programmes covering a minimum of six and a half years during this period.[4]

## Changing emphases in traditional conditionality

Important changes have taken place in the content of IMF conditionality in recent years, but these have not incorporated any fundamental movement on the provisions that have traditionally formed the bedrock of Fund programmes: restriction of domestic credit creation and budget deficits, and currency devaluation. The financial programming model used in the estimation of country performance criteria has remained largely unchanged for many years. It has not, for example, been modified to bring in more of the supply side of the economy, despite modelling advances in this area (see, for example, Khan and Montiel, 1989). Even in the longer-term, more production-oriented ESAF programmes, requirements (often in the form of preconditions) for currency devaluation and traditional performance criteria on fiscal and monetary variables remain at the centre of the policy package.[5] Indeed, insistence on devaluations has increased over the years. Polak (1991: 36) calculates that in the period 1963–72, 32 per cent of all programmes (excluding currency-union countries) incorporated exchange-rate measures; that this proportion rose to 59 per cent during 1973–80;[6] went up again to 82 per cent during 1981–3 and rose to 'nearly 100 per cent' thereafter. The IMF was heavily engaged in the 1994 decision to devalue the CFA franc from a rate against the French franc which had remained unchanged since 1948, an event which triggered a rapid increase in its lending to Franc Zone member countries.

While there has been no retreat on the eternal verities of Fund conditionality, there have been significant changes of emphasis. One change evident in recent staff writings is the insistence that the policy coverage of programmes should be broad, and well-balanced as between its traditional demand-management provisions and more supply-side measures. In language radically different from what might have been expected in an equivalent document ten years earlier, a 1989 staff paper complained that governments often preferred to stick to the familiar parameters of Fund programmes, leaving a risk that inadequate attention to structural weaknesses could endanger or slow down adjustment. It went on to raise the question of 'how far the Fund staff should go in making key structural adjustments a prerequisite for granting an arrangement, or for interrupting purchases'. Other internal papers echo this theme, as does the 1991 *Annual Report* (p. 24): 'A recurring theme of discussions with developing countries . . . was the need to sustain broadly based macroeconomic stabilization and structural reforms over the medium term . . . '.

Another shift has been an upgrading of the objective of trade liberalisation. While the desirability of this has long been written into Letters of Intent, it was seldom taken seriously in the 1960s and 1970s[7] and our earlier investigations showed little association between Fund programmes and actual liberalisation (*The Quest*, pp. 236–8). However, we have already drawn attention to the 1989 document which, unusually, states as the main objective of stand-by programmes a sustainable balance of payments 'in the context of an open trade and payments system'. This was no casual insertion. Internal evidence indicates that some countries have been refused programmes because of unwillingness to act in this area; and liberalisation is now sometimes written in as a performance criterion.

There have also been important changes with respect to the Fund's thinking on fiscal conditionality, particularly under the influence of the head of its Fiscal Affairs Department, Vito Tanzi.[8] Internal staff papers now write disparagingly of the past tendency of programmes to go for 'quick fixes' and to concentrate too much on aggregate spending ceilings. In consequence, Fund missions now pay more attention to the details of the budget and to how governments are going to secure the overall spending cuts they promise.[9]

A number of influences have pushed the Fund in this direction.

For one thing, governments have become increasingly canny in evading overall spending ceilings; for all their experience, it is difficult for Fund missions to compensate for informational asymmetries and close all possible loopholes. Secondly, the upgrading of the growth objective, reported above, has highlighted the obvious fact that the way the burden of cuts is spread across the government's expenditures can make a great deal of difference to their impact on the economy. In the absence of safeguards, there is, for example, a near universal tendency for governments to cut first on capital expenditures, and then on maintenance and current supplies. They may also be tempted to impose disproportionate reductions in the provision of economic services, in attempts to avoid reducing the size of the civil service and/or the military budget. To put it differently, there is often a trade-off between the quantity and 'quality' of government spending reductions, with given balance-of-payments and growth targets being consistent with smaller high-quality expenditure cuts (which protect productive expenditures) or larger but lower-quality reductions.

Fund teams therefore nowadays involve themselves more in the details of budget making. Tanzi, writing in 1989, complained that this had made only a limited difference, partly because of political resistances, and that spending cuts still fell particularly hard on government capital formation (Tanzi, 1989: 25–8). However, Fund staff report that the resistances have weakened and that stipulations relating to the quality of fiscal-deficit reduction are increasingly being written into ESAF conditionality. Nevertheless, findings reported in Chapter 3 suggest that the Fund's measurable impact on fiscal outcomes in programme countries remains quite restricted.

Another new emphasis, and often a *de facto* extension of conditionality, is the far larger volume of technical assistance which is now provided by the Fund. This was formerly a rather minor part of its work but has been expanded rapidly in recent years, frequently in support of a stand-by or other high-conditionality programme.[10] Acceptance of such assistance is commonly made a condition for agreeing a programme and the Fund has used the personnel thus installed to improve the execution and monitoring of programme measures, especially in the area of fiscal policy.

## The rise of structural conditionality

The spread of the Fund's influence beyond its traditional remit of fiscal–monetary matters into 'structural' matters has been rather more dramatic. Although it was a process begun in the 1970s with the introduction of the EFF, it has been taken a good deal further through SAF and ESAF programmes, and through the resumption of EFF lending. Even some stand-bys nowadays contain structural provisions, for example with regard to pricing policies, although only exceptionally are these preconditions or performance criteria.[11] Examination of the content of this 'structural' conditionality indicates three main but overlapping thrusts: (a) to increase the role of markets and private enterprises relative to the public sector, and to improve incentive structures; (b) to improve the efficiency of the public sector; and (c) to mobilise additional domestic resources.

As regards the first of these, programmes commonly contain measures to bring prices closer to market-clearing levels, for example by the reduction or elimination of public subsidies and the removal of price controls. The pricing of petroleum products has been a frequent target in countries which have subsidised this or followed tax policies which kept final prices below true economic costs. Agricultural producer prices have also featured quite often, to ensure that adequate incentives are offered. The operation of the market mechanism has also been promoted through the liberalisation of trade and payments, already mentioned; and the realm of the private relative to the public sector has been extended in a substantial number of programmes by requirements for the privatisation of public enterprises.

Privatisation shades into the second chief thrust of 'structural' programmes – raising the economic efficiency of the public sector. In addition to sales of enterprises to private shareholders, programmes have provided for some of the more chronically inefficient public enterprises to go into liquidation. More often, there have been provisions for diagnostic studies of ailing enterprises and the preparation of action programmes, which have led in some cases to financial and/or managerial restructurings and other rehabilitation measures (not necessarily as a preliminary to privatisation). Some programmes have incorporated the device of performance contracts for public enterprises; more have required overhauls of enterprise pricing policies.

A fairly wide range of fiscal reforms have also been promoted as the means of improving public sector efficiency. These have included tax reforms intended to reduce incentive distortions and to increase the buoyancy or elasticity of the tax system, including improvements in revenue collection. As mentioned above, much detailed attention has also been paid to improving the content of government expenditures. In quite a number of cases this has included reviews of, and/or reductions in, the size of the civil service, and examination of its remuneration levels and incentive structures. Attempts have been made to protect capital formation in government budgets, although it is doubtful whether these have been effective.

Strengthening the revenue side of the public finances also promotes the third thrust mentioned above: improved domestic resource mobilisation. In addition, a substantial proportion of programmes have included provisions for the reform and strengthening of the financial sector. These have included the recapitalisation of banking institutions and restructuring their asset portfolios; the decontrol or reform of interest rates, generally with a view to raising these to positive real levels; the development of new financial instruments to encourage the development of capital markets; and actions to strengthen the supervisory and technical assistance capabilities of central banks.

Some indication of the relative importance attached by the Fund to these various provisions is provided by information on the subject coverage of benchmarks and performance criteria in SAF and ESAF programmes (although this is also influenced by the feasibility of defining measures with sufficient precision for them to be given this status).[12] On this basis and paying special attention to the more rigorous ESAF programmes, fiscal reforms were found to have been given the greatest weight by a substantial margin, with reforms in the areas of public enterprises, the financial sector, agriculture and trade, scoring about equally in second place. In about two-fifths of programmes, progress on one or more of these 'structural' provisions was given the status of a performance criterion.

It is clear, then, that the IMF's 'structural' programmes have substantially widened the coverage of its conditionality. It has responded in this way to earlier criticisms that it did too little to address balance-of-payments weaknesses emanating from the productive system and weaknesses in economic structures. It is also clear, however, that the provisions just described take a somewhat partial view of what constitutes 'structural'. It is a view which coincides rather precisely with the pro-market, disengage-the-state approach of its major shareholder countries, and this fact doubtless smoothed the way for the Fund to broaden its conditionality in this way.

A further aspect of the IMF's shift into 'structural' programmes through the vehicles of the SAF and ESAF is that it has introduced a potentially important modification to programme-formulation and negotiation processes. As described earlier, SAF and ESAF programmes are intended to be based on three-year Policy Framework Papers drafted jointly by the borrowing government, the IMF and World Bank. In the early years of this innovation it was somewhat notoriously the case that in reality the role of the government in the process was often minimal, and that the PFPs were written in Washington by Fund and Bank staff. As much was admitted by a former senior member of the Fund's staff dealing with Africa, writing of the mid- and later 1980s: 'Unfortunately the PFP so far has been primarily a subject of negotiation between the staffs of the Fund and the Bank' (Goreux, 1989: 162). He might have added, 'with the Fund generally having the final say'.

However, it appears that this situation has gradually improved and that some borrowing governments have acquired more of a voice in the process. There has been some relaxation in Fund procedures which has given its missions more freedom to modify draft PFPs to accommodate government views. And to counter the long-standing criticism that IMF missions largely confine their discussions to the Finance Ministry and Central Bank, there has also been an effort to involve a wider range of ministries in the preparation of these documents, including the line ministries which have to implement the adjustment measures agreed. Yet a further widening of the process is occurring through greater consultation with bilateral and other aid donors, both to be able to take their views into account and as a way of mobilising their financial support for the programmes. In consequence, the Fund is on record as believing that the PFP process 'has become an increasingly effective instrument for designing medium-term policy measures' (IMF, *Annual Report, 1991*: 57), although there are doubtless a good many governments which still see the PFP as essentially a 'Washington' document in which their views and priorities are inadequately reflected.

## IMF Flexibility

In the past one of the commonest complaints about the Fund, particularly where governments were seeking to stabilise their economies in the face of economic and political turbulence, was that it was too rigid in its negotiating stances, too uniform in its approach to programme design. Has this too changed during the last decade? To answer this we can draw on the survey of 17 country case studies described in more detail in Chapter 3. Although these provide only limited information, what is remarkable about them is the relative absence of complaints about inflexibility.

Matin's (1986, 1990) writings on Bangladesh's programmes are an exception, arguing the need for a more judgemental approach to IMF decisions about continuing access to credit, rather than a mechanistic approach based on the observance of quantified performance criteria. Information on the fiscal effects of exogenous shocks also suggested that the Fund was on occasion too reluctant to adjust performance criteria in the light of changed circumstances, leading to programme breakdowns. Our materials also include many cases in which programmes were suspended or abandoned, and some of these breakdowns may have been the result of Fund rigidity. However, if we confine ourselves to the literature on the countries surveyed, there are rather more examples of apparent flexibility than of the opposite.[13]

We should, however, make a distinction between the type of flexibility which leads the Fund to avoid taking up a rigid position during programme negotiations and flexibility in their execution. As regards the former, the Fund could be described as having adopted relatively amenable positions in several of the countries studied. A stand-by programme negotiated with Mexico in 1986 is of special interest in this connection because it incorporated elements of pliability in a programme which was unique. It contained provisions that varied the terms of the programme depending on external trading and other conditions, with automatic modification both of the performance criteria and of the size of the credit according to the value of petroleum export earnings and the growth of the economy. Martin (1991, Chapter 2) reports little equivalent pliability by the Fund in its dealings with African debtor countries. In his view, 'Even debtor-negotiators who were well-prepared, united, flexible, experienced, persuasive, skilled in politics and economics, and who understood the Fund, received few concessions' in programme negotiations (p. 42). Some IMF missions are described as having a dogmatic 'take it or leave it' attitude; even sympathetic staff apparently regarded themselves as having little freedom to depart from their briefs.

Martin's findings become consistent with others showing greater Fund negotiating flexibility, if we bear in mind the political isolation, or insignificance, of many African debtor countries. For country experiences point clearly to a strong correlation between the IMF's negotiating stance and political pressures that may work upon it to 'do a deal' with the country in question.[14] This political favouritism is reported more fully in the following chapter. In addition to clarifying differences of treatment as between African and major Latin American debtor countries, that political favouritism is an important factor is implicitly supported by Stiles's (1991) findings linking Fund flexibility with the extent of personal involvement of the Managing Director, which is strongly related to the extent of politicisation of the decision. Similar political influences also underlie Martin's finding that the Fund secured relatively few of its objectives in negotiations with Sudan and Zaïre, with a staff member describing these programmes as 'cobbled together so a few donors could justify propping up their friends' (Martin, 1991: 53).

It is tempting to conclude from this that Fund missions display negotiating flexibility only when the institution comes under external pressures to do a deal, but this may be unfair. There is internal evidence, for example, that the IMF gives more weight than formerly to the political sensitivities of its task, with a 1989 staff report observing:

> programs need to be developed with due regard to the domestic social and political objectives, the economic priorities, and the circumstances of the members. In framing their economic programs, member country authorities may be faced with important constraints that limit their ability to undertake as bold a program as might be called for on purely economic and technical grounds. In such circumstances, an approach that might be second best in an economic sense may be accepted, provided that it has sufficient focus on external viability.

Overt acceptance that political conditions in the borrowing country may require the Fund to settle for a sub-optimal programme is new language in the staff papers we have seen over the years, although there was doubtless *de facto* acceptance of this reality in many earlier programmes. We should, moreover, not exaggerate this acceptance, for only four of the 40 programmes reviewed in the 1989 staff report were actually classified as second-best in the sense just described.

A final aspect of flexibility in negotiation on which there has been a little progress has been in the direction of including contingency provisions – clauses which vary the terms of a programme and/or the amount of supporting finance in the light of unforeseen changes in external conditions. The 1986 Mexican agreement mentioned above excited hopes that such provisions would be extended to other countries. Had this happened it would have been a major movement in the direction of greater flexibility, but in the event Mexico was treated as a special case; the politics was paramount again.

The extension of the old Compensatory Financing Facility to include additional finance in the event of certain unfavourable developments beyond the control of borrowing governments, described earlier, was another potentially useful move, but the terms under which this support becomes available to governments are unattractive and the new provision has been little used. Of greater practical value has been a relaxation of what the Fund calls 'access policy', i.e. the rules governing the amount of credit it may grant in support of a given programme. These are now sufficiently loose to permit credits to be enlarged, should external conditions turn against a borrowing country and threaten to derail the programme. A substantial number of countries have benefited in this way.

Other changes in IMF practices to take greater account of the effects of exogenous developments have been far less favourable to borrowing governments. One is the more frequent inclusion of contingency provisions which increase the stringency of credit ceilings when *favourable* external developments occur, e.g. larger-than-forecast aid receipts. Another is requirements in programmes for larger increases in external reserves, *and therefore more stringent policies*, against the possibility of adverse external movements.

When we turn from flexibility in negotiation to flexibility in implementation there are additional developments to describe. The most important change in the 1980s was the adoption as more or less standard practice in both stand-by and medium-term programmes of provisions for six-monthly review missions. The purpose of these is to assess progress with the programme over the preceding half-year and to reconsider or determine performance criteria and other provisions for the following half-year. As such, they reduce the Fund's reliance on predetermined quantitative indicators and facilitate the modification of programmes in the light of changing circumstances. They therefore should reduce the risk complained of by Matin (above) of mechanistic reliance on quantified credit ceilings and the like. They also make it easier for the Fund to incorporate provisions which cannot be reduced to quantified indicators, and situations where the database is poor and uncertainties are large.

There are costs, however, because the more judgemental nature of the review mission process increases opportunities for political favouritism, and governments have less assurance of continued access to an agreed credit. For this latter reason, the doyen of the Executive Board suggested that borrowing governments should be given a choice between programmes based around performance criteria and programmes containing only periodic reviews (Kafka, 1991).

## AN ASSESSMENT

### How much has the Fund really changed?

A good many of the changes described above are certainly to be welcomed. The broadening of programme objectives to give greater weight to growth and to be more sensitive to the impact of adjustment measures on vulnerable groups is a change that moves the Fund in directions that have long been urged upon it by developing countries and independent observers. Although more controversial, the broadening of programmes to give them more structural and 'real economy' coverage may also be seen positively, particularly when viewed in conjunction with the lengthening of the Fund's time horizons, for it has increased the appropriateness of the institution's conditionality and financial support. This is particularly so for its poorer, less developed member states. For them, acceptance of the need for special forms of support, formally enshrined in the SAF and ESAF, was also an important move forward, and we have suggested that the PFP mechanism associated with these facilities has been evolving in ways which increase the involvement of borrowing governments in programme formulation. We should further welcome the apparent increase in the scope for IMF flexibility over programme implementation resulting from the greater use of review missions, and its greater freedom to increase the amount of credit when necessitated by deteriorating external circumstances.

All these movements are in desirable directions because they enhance the Fund's ability to provide effective balance-of-payments assistance to its developing country membership. What is much more difficult to judge is *how far* these changes have actually taken the agency – how much difference have they made at the country level? Inevitably, there is a gap between the policy statements of the Managing Director and the actuality on the ground, but is it an unbridgeable chasm?

We should here recall from Table 2.2 that, while the proportion has fallen sharply in recent years, a sizeable amount of IMF lending to developing countries is still in support of stand-by programmes. By common consent, these have changed only modestly over the years. They are still dominated by the Fund's traditional balance-of-payments objective. They still embody the same financial programming approach and performance criteria that have long been the hallmark of stand-bys, and continue to take a demand-restraint-plus-devaluation approach to balance-of-payments management.

The sharp increase in the share of lending in the form of EFF and ESAF credits has placed the Fund in a better position to pursue a growth objective. The fact that growth features more prominently in public statements of their objectives, that they are less short term in nature, addressed more to supply-side constraints and commonly accompanied by World Bank 'structural adjustment' programmes, should mean that ESAF and EFF programmes are associated with higher levels of economic activity. The revival of EFF credits shown in Table 2.1 should be a positive move, for evidence from the 1970s indicated that EFFs were more successful in protecting economic growth than stand-bys (*The Quest*, pp. 248–9), although they achieved little success in strengthening countries' balances of payments. But the step is too recent for there to be reliable evidence yet on the results achieved.

As for the announced growth orientation of ESAF programmes, this appears to be mostly a matter of public relations. The following chapter shows that in the Fund's first (1993) evaluation of its experiences with ESAF programmes, its assessment was overwhelmingly based on progress towards external viability. The growth record received so little attention as to imply a rejection of the relevance of the growth criterion. It is similarly doubtful whether the apparent upgrading of the growth objective has made much difference to stand-bys. Growth is still not accepted as constraining the design of stand-by programmes, apart from a greater reluctance to submit to the Executive Board programmes which are manifestly inconsistent with a resumption of growth.

A small amount of evidence on the extent of change in stand-bys can be gleaned from the few recent stand-by programmes which have been made publicly available by the borrowing governments. We have information on stand-bys in Argentina in 1990, in the Dominican Republic in 1991 and in Jamaica in 1991. All reflect some, but not all, of the conditionality trends discussed above.

The *Argentinian* programme had considerably more 'structural' content than would have been expected a few years earlier. This included provisions for the reform and/or privatisation of certain public enterprises, the liberalisation of interest rates and various tax reforms. Performance criteria were, however, restricted to the types of variables that have long been conventional in stand-bys, and there were no less than eight of them. There was nothing in the programme which could justify calling it growth-oriented, and it was silent on the protection of poverty groups.

The programme in the *Dominican Republic* differed in that the government explicitly acknowledged its responsibility to protect the living conditions of the poorest and committed itself to the creation of a 'social fund' and promotion of labour-intensive social services. However, here too there was little to protect economic growth. There was a commitment to measures intended to boost domestic saving but, against this, the programme envisaged a disproportionately large cut in government capital formation. There was little in it of a specifically 'structural' nature.

The *Jamaican* programme incorporated further variations. Although it was unspecific about growth, it was set in a medium-term framework and envisaged an increase in government investment expenditures, both absolutely and proportionately. Although it is not clear that this was a negotiated provision, the programme accommodated a rise in poverty-related state spending and specific poverty-alleviation measures. The overtly 'structural' content was limited but included actions to reform certain public enterprises, the liberalisation of import marketing and the elimination of some price controls.

There are similar doubts about the extent to which movements in Fund policy have actually protected poverty groups from adverse programme effects. The evidence is reviewed in the next chapter but, while there are specific instances where safety nets have been erected (particularly in the formerly communist countries of Eastern and Central Europe), it can fairly confidently be asserted that, taking an overall view, rather little has actually been achieved thus far. Indeed, in an interview Managing Director Camdessus expressed dissatisfaction with the operational impact of the apparent policy shift. Part of the problem, of course, is that a good many borrowing governments do not themselves give high priority to protecting the poor.

## A summing-up

If one steps back and asks the question, just how different is the Fund today from how it was 10 or 15 years ago?, different well-placed observers reach contrasting conclusions. A former long-standing senior member of the Fund's staff who is now a member of its Executive Board sees the recent changes as 'less a break than an elaboration of earlier policies' (Mohammed, 1991: 245), whereas Polak sees their cumulative effect as having wrought 'a fundamental shift' in the Fund (1991: 2). Our own assessment stands in an intermediate position. It does less than justice to the extent of change on the IMF's stance relative to its developing country members to dismiss it as merely an elaboration of past approaches, even though there are admittedly historical antecedents for many of the shifts. But to say that the Fund's relations with its Third World members have changed fundamentally is to over-egg the pudding. The policy changes described in this chapter have been in a generally desirable direction and are making a difference to programme realities. But in many cases the difference is modest – a good deal smaller than the extensive changes in economic policy stances of many of the Fund's developing country members over the same period.

The extent of difference, in any case, depends on the type of developing country we are considering. The changes have been largest in programmes in African and certain other low-income countries, which have been able to benefit from the SAF/ESAF innovations. They have often been quite small in middle-income developing countries, which are not eligible for SAF/ESAF and which more often receive assistance in the form of short-term stand-bys. That large and important group of countries is unlikely to consider that there has been a fundamental shift. One of the problems now confronting the Fund is to develop ways of reducing the short-term, demand-management bias in its dealings with these middle-income countries. The revival of the EFF can be understood in this light, but it is at present only a very partial response, with a mere three EFF programmes in middle-income developing countries as at March 1994 (Table 2.2).

In drawing up the balance sheet, we should also remember that the more important reforms described earlier have been achieved at a cost which many developing country governments will regret: a proliferation of conditionality. This is most obvious in the case of the ESAF and EFF; we have already described the wider scope of the policy coverage of these programmes. It has also occurred in some degree in stand-bys. In addition, there has been extensive use of preconditions in SAF/ESAF programmes. However, the starkest evidence on the proliferation of conditionality is provided by the following averages for the number of performance criteria per programme (Polak, 1991: 14):[15] 1968–77, below six; 1974–84, seven; 1984–7, nine and a half. Evidently the attempt to use the 1979 conditionality Guidelines to limit the number of performance criteria was unsuccessful.

This proliferation (or 'deepening', as the staff would describe it) is a problem for the Fund itself as well as for implementing governments. For one thing, it increases the risk of slippage in programme execution and may help to explain the rising proportion of programmes breaking down in recent years (see Chapter 3). It also seems that the particularly stringent conditionality attached to ESAF programmes has had a strong deterrent effect on the utilisation of this facility, which explains the failure for lending from this facility to grow as expected after its introduction, as shown in Table 2.1. As Feinberg (1991: 6) ironically observed, 'When the ESAF was initiated, Fund management emphasised that the resources would only be used to support strong adjustment programmes. Evidently the Fund has honoured this commitment.' It may be, on the other hand, that a recent scaling-down in the number of preconditions in ESAFs is an attempt to encourage more governments to use this facility.

# 3

# PROGRAMME EFFECTS

## What can we know?

Despite the changes described in the previous chapter, controversies about the appropriateness and effects of IMF stabilisation programmes remain lively and have generated a substantial literature. For different reasons, both the Fund and its critics like to believe that its programmes[1] have powerful effects, although they disagree over whether these are benign or malign. On the basis of evidence relating mainly to the 1970s, on the other hand, *The Quest* (Chapter 7) suggested that the controversy was a case of much ado about nothing: that the effects were relatively slight in either direction.

This uncertainty is unsatisfactory. At any one time a large number of developing countries have IMF programmes in place, in the negotiation of which large amounts of scarce skills and information have been deployed by both parties. The macroeconomic policies of borrowing countries are strongly influenced, not to say dominated, by the terms of the programme. Virtually all the Fund's financial resources and a large proportion of its staff are devoted to these programmes. Debt reschedulings are usually conditional on the debtor government signing a programme agreement with the Fund. Credit and investment decisions, by both public and private agencies, are also liable to be influenced by whether or not a Fund programme is in place. It would be absurd to place so much weight on the existence of an IMF programme if, in fact, they had little effect. At the same time, as we shall see, there are major difficulties in the way of forming a definitive judgement, leaving questions about how much it is feasible to know about programme effects.

The purposes of this chapter are to explain the methodological problems, survey the state of existing knowledge on programme effects, and to present new evidence.

## METHODOLOGICAL ISSUES

### Pitfalls

Formidable obstacles stand in the way of arriving at firm evidence on the effects of IMF programmes:

(a) Although there are econometric techniques for this purpose, it is difficult in practice to disentangle programme effects from the many *other influences* which bear upon the performance of an economy.

(b) It is difficult to distinguish between the effects of the policies in Fund programmes and the effects of the *finance* which accompanies the programmes. This may be a particularly large difficulty with credits linked to World Bank structural adjustment lending (e.g. ESAF programmes), because large volumes of finance are sometimes associated with such policy packages. In these cases evaluating the effects of the programme of either institution is somewhat artificial and it may be preferable to study their joint effects. However, this has not been attempted in the literature. Similar problems arise where Fund credits are linked to major debt relief operations, for debt relief is equivalent to additional finance.

(c) There are difficulties in selecting adequate *performance indicators* in statistically based cross-section analyses. This is perhaps most acute in the case of the balance of payments (BoP), the improvement of which is the chief objective of IMF programmes. The overall balance (i.e. the balance on monetary account with opposite sign) is the most commonly used indicator, but this is vulnerable to the difficulties outlined in (a) above and to the influence of any capital inflows and debt relief triggered by agreement with the Fund. The current account balance is another popular indicator, but the difficulty with this is that governments facing a foreign-exchange crisis with exhausted reserves have to find ways of limiting the current account deficit to whatever money is expected to be available to finance it. They often do so by means of severe import cuts, which are liable to have adverse effects on economic performance, including ability to export,[2] so that a reduction in the current account deficit does not necessarily tell us anything about the underlying strength of the BoP situation. Use of the overall balance is also affected by this limitation. The essential difficulty is that progress on the BoP cannot adequately be monitored by reference to one or two residual balances. To a lesser extent, similar problems arise with other performance indicators commonly used.

(d) There are problems also with the *period of analysis*, with results often sensitive to the choice made. Should the impact of Fund programmes be assessed only for the period of the programmes or for some longer period? The time-lags between changes in policy variables such as the exchange rate or domestic credit and BoP results suggest a more extended period, but the longer this is, the louder becomes the extraneous 'noise' in the tests. The period of analysis also creates difficulties in comparisons of countries with and without IMF programmes: given that programmes are agreed at different times and for varying periods, it is not clear over what period the performance of non-programme countries should be measured. (For an example of this problem, and of the sensitivity of results to the period chosen, see Gylfason, 1987.)

(e) The rigour of policy conditionality varies according to *programme type*. Only rather relaxed conditionality is attached to first-tranche stand-by and SAF credits, as compared with upper-tranche credits. ESAF conditionality is the most demanding of all (see Appendix to Chapter 1). Even among programmes of the same type there are considerable variations. Presumably results should differ in extent and speed according to the severity of conditionality, but multi-country tests invariably treat all programmes as equal.

(f) Relatedly, some programmes are more fully implemented than others, so effectiveness tests should ideally adjust for the *degree of implementation*. Indeed, as we shall see on pp. 58–66, a large proportion of programmes break down in the course of their intended lifetime.

Above all, however, methodological discussions are dominated by the *problem of the counterfactual*, taking as the central question whether Fund programmes result in a better situation than would obtain in their absence. Khan (1990: 198) puts the point eloquently:

> The counterfactual is perhaps the most appealing yardstick against which to assess program performance and the standard most widely employed in economics to measure the impact of government policy interventions. What would have happened in the absence of a Fund-supported program is by no means the only standard against which to judge the outcome of programs, but in many cases it is the most appropriate one. However, the counterfactual cannot, by definition, be observed and must be estimated or approximated. *The various approaches used in evaluation studies should thus be judged in terms of how good they are in providing estimates of the counterfactual* [emphasis added].

There are three, linked, principal reasons why it is desirable to use the counterfactual yardstick. The first is the importance of disentangling the (generally adverse) effects of the state of the economy – often a crisis situation – immediately prior to programme adoption from the effects of the programme itself. Some critics of the Fund skate over this and illegitimately impute to the programme adverse outcomes which are more appropriately attributed to the initial situation. If an economy starts with a foreign-exchange crisis, rapid inflation, major supply-side bottlenecks and large-scale excess demand, it is scarcely surprising if in the following two or three years there is little growth, consumption standards fall and investment stagnates. This tendency to attribute to programmes an economic deterioration which is predetermined by the initial situation is compounded by the fact that by far the easiest, and most commonly employed, type of test to apply in cross-country tests is the comparison of economic variables before and after programme introduction – the so-called before–after test.

A second reason for applying a counterfactual test is that programmes are commonly knocked off course by shocks beyond the control of governments and the Fund. Abnormal weather, organised violence, changing conditions in world capital markets are examples, but unexpected changes in borrowing countries' barter terms of trade are probably the most common and important. Given continuing turbulence in world economic conditions, it is inevitable that some programmes will 'fail' because of such factors.[3] By definition, the counterfactual would be no less affected by these shocks, so comparisons between this and programme results would eliminate the bias.

This leads to the third reason why the counterfactual problem is regarded as so central: that this literature has a strongly normative content, treated as evidence on the utility of the IMF as an institution. This is unfortunate, if inevitable. Since both exogenous shocks and non-implementation are reasons for programme ineffectiveness which cannot be laid at the door of the Fund, it would be desirable to separate the 'positive' assessment of programme effects from normative evaluations of the institution, and knowledge of the counterfactual would facilitate this separation.

## A menu-of-information approach

How should researchers respond to these difficulties, particularly to the unknowability of the counterfactual? Although the problem of the counterfactual must indeed be taken seriously, it dominates the literature unduly. In particular, it is wrong to use their adequacy in proxying the counterfactual as the exclusive test of the various techniques because, as Khan noted, other criteria are relevant too.

Take before–after tests, for example. They do not cope well with the counterfactual problem, but they do still provide useful 'positive' information. It is perfectly sensible to ask, did this programme lead to an improvement on the initial situation? *It was almost certainly in order to achieve this that the government negotiated a Fund credit.* The Fund's Articles state that its credits are intended to give governments an opportunity to 'correct maladjustments in their balance of payments', which implies that programmes are intended to improve on the initial BoP situation. To the extent that the situation is not improved, we have to examine the reasons for the shortfalls. These may or may not lie within the responsibility of the IMF, but we shall have gained useful information.

More broadly, the position taken here is that the various tests of programme effects each provide useful, although limited, information, and we should therefore use them all in order to build up a menu of 'positive' information. We must, however, be careful about using these results for any normative assessment of Fund effectiveness. In this spirit, what follows is a brief résumé of the positive information that the various types of tests can yield, and their chief weaknesses, summarised in Table 3.1. Using this approach, we can build up an inventory of what we can (and cannot) know about programme effects:[4]

(a) *Before–after tests*[5] discussed above. These have the advantage of being relatively easy to conduct. They yield information on whether programmes were associated with an improvement on the initial situation. They are also useful for information on whether programme effects are sustained, and whether results are affected by country or programme type. Results can be misleading, however, because of the initial situation and exogenous shocks, and are easily misused in normative discussions.

(b) *Target–actual tests.*[6] These compare outcomes with programme targets, giving an indication of the extent to which intended results are achieved. However, such quantified targets are no better than the models which generate them, so discrepancies between targets and actualities may be a result of poorly performing models rather than unsatisfactory economic responses. Like the before–after approach, this method does not provide a proxy for the counterfactual and yields no information on the costs of achieving results. It might give unduly negative results because of unrealism in programme targets; but that the targets may be unrealistic is itself useful information. Its main operational snag is that it requires information on programme targets which is usually not publicly available. For multi-country studies, its use is confined to IMF staff (various unpublished staff studies use this approach) or researchers to whom the Fund is willing to release the information.[7]

(c) *With–without tests.* (For examples see Donovan, 1982; Loxley, 1984; Goldstein and Montiel, 1986; and Gylfason, 1987.) These compare a sample of programme countries with a control group of non-programme countries. This is an approach to the counterfactual question, using the experiences of the control group as a proxy for what would otherwise have happened in the programme countries. The difficulty with this approach is to achieve a truly comparable control group. Although later attempts (e.g. Goldstein and Montiel, 1986; Gylfason, 1987) make major efforts to ensure this, it is intrinsically difficult because the decision by control countries with equally severe BoP problems not to adopt a Fund programme can be expected to be of large policy significance, vitiating the comparability of the two samples.[8]

(d) *Comparison of simulations.*[9] These are based on policy models of one or a group of economies. They can be used to predict outcomes when IMF-type policy actions are introduced and to compare with a counterfactual. Depending on model specification, they can correct for the effects of exogenous variables. Their usefulness depends crucially on the appropriateness of the model used, with the danger that results reflect model specification rather than reality. Different models can produce substantially varying results. Also, this technique does not yield direct information on programme effects, *per se*, and it assumes that model parameter values would be unchanged by the policies adopted.

(e) *Generalised evaluation.*[10] This proceeds by building a model in which various policy and other explanatory variables are incorporated in regression equations which take the performance of the BoP, inflation and economic growth as dependent variables and in which a dummy variable is included to catch the influence of any Fund programme. This approach thus offers a direct test of programme impact, although its inclusion simply as a dummy means that it cannot provide refined information about this. Here again, the value of the method is model-dependent.

(f) *Extent of programme completion.* This provides some information on the extent of programme implementation, because governments are refused access to credit tranches when they fail to comply with programme performance criteria. As such, it throws light on an area which most other tests neglect: the reasons for programme failures. Once the basic information has been collected, it can readily be classified so as to explore results for different programme and country types. The study of programme non-completions (pp. 58–66) sets out the results of such an exercise – the only cross-country example of which we are aware.

(g) *Country or programme case studies.* In principle, the case-study approach can avoid the various disadvantages adduced above and can go much further in explaining the results obtained. Its abiding disadvantage is that it does not yield results that can be generalised across countries. Indeed, it can be destructive of the desire to infer general conclusions by emphasising the uniqueness of each case. A solution can be sought by undertaking a number of case studies with a common methodology but this demands a lot of resources.[11] An alternative approach is reported in the survey presented on pp. 86–119. This is a comparative study of the results of a large number of already published individual case studies, although it is subject to the limitation that these were not devised within a common analytical framework.

Table 3.1 summarises the types of information that the various methods of evaluation are able to yield, and their chief weaknesses. While some are more informative than others, each method has its uses and by combining them we can build up a substantial menu of 'positive' information about programme effects, which is the main task of this chapter. Before moving to this, however, we should note the inability of most multi-country studies to go far in *explaining* the results obtained, as indicated in the lower part of Table 3.1 (items 8–11). If we are concerned to understand why programmes are sometimes unable to achieve their objectives, the table suggests four lines of explanation: the influence of the initial situation and of exogenous shocks; incomplete programme execution; inadequate programme impact on the intended instrument variables; and inadequate financial support for the programmes. To these can be added a fifth: possible defects in programme design. While certain of the multi-country approaches can attempt to screen out the effects of the initial situation and some exogenous shocks, they throw little light on the other factors. This is an area in which the case-study approach scores especially heavily.

Given the methodological pitfalls, the question arises of how much we can know about the effects of Fund programmes. The remainder of this chapter is devoted to the results of empirical tests, and their implications, beginning with a survey of published studies.

## THE EXISTING LITERATURE

We concentrate here mainly on the statistically significant results of the various studies, but not to the exclusion of other suggestive results which may fail the 95 per cent minimum significance level or for which statistical significance was not calculated. We also concentrate chiefly on published studies, although some reference is made to unpublished IMF staff studies. No attempt will be made to summarise systematically the results of each study. Our purpose is rather to paint a composite picture of what can reasonably be inferred from these studies taken together. Since we wish this to be

*Table 3.1* Uses and limitations of alternative tests of IMF programmes

| | *Before–after* | *Target–actual* | *With–without* | *Comparison of simulations* | *Generalised evaluation* | *Completed–uncompleted* | *Case study* |
|---|---|---|---|---|---|---|---|
| *Programme results* | | | | | | | |
| 1. Does the programme improve on the initial situation? | Y | Y | Y | | Y | | Y |
| 2. Do programme countries do better than non-programme? | | | Y | Y | Y | | |
| 3. Do programmes improve upon likely alternative outcomes? | W | W | D | Y | Y | W | Y |
| 4. Are programme effects sustained? | Y | | Y | | Y | Y | Y |
| 5. At what costs are their results secured? | | W | | | Y | | Y |
| 6. Can the results be generalised? | | | D | D | Y | | W |
| 7. Do results differ for: | | | | | | | |
| (a) country types? | Y | | Y | | | Y | |
| (b) programme types? | Y | | Y | | | Y | |
| *Result determinants* | | | | | | | |
| 8. Influence of exogenous factors | W | | | Y | Y | | Y |
| 9. Programme implementation | | | | | W | Y | Y |
| 10. Impact on instrument variables | | Y | | | | | Y |
| 11. Impact on financial flows | | | | | | | Y |

*Key*: Y = The test provides useful information bearing on the issue in the left-hand column.
W = The test is particularly weak in this area.
D = Debatable. There is disagreement in the literature on whether this test can provide useful information of the type asked for.

relevant to the contemporary situation, special weight is attached to tests which include data for the 1980s and 1990s. The discussion will follow the sequence set out in Table 3.1. However, we shall not follow the customary procedure of treating GDP growth as a target variable, bringing it in rather in the discussion of the *costs* of adjustment.[12] We take the BoP indicators as the chief target variables, with reductions in inflation as a second-order goal.

### Change on the initial situation

Our first question, then, is whether Fund programmes result in an improvement on the initial situation. The results of before–after tests are the most pertinent here. Generally speaking, the results of these are not encouraging. None shows a significant improvement in the BoP current account. The results are more mixed on the overall balance but, of the general evaluations, only the most recent, utilising data for 1973–88, shows a significant positive effect (Khan, 1990: Table 2). Pastor (1987) on 1965–81 data obtains a similar result but his tests were confined to Latin American countries. None of the other before–after tests yields a significant impact on either BoP indicator and in some cases the sign is 'wrong'. The ability of programmes to reduce inflation appears even weaker, with results showing the inflation rate to increase as often as it reduces but a virtual absence of statistical significance.[13] If governments go to the Fund in order to improve on the initial situation and if we confine ourselves to significant results, it seems that its programmes are often unable to achieve the desired turn-around, although Khan's positive result is the only one which includes much recent information.

### Comparisons with targets

A rather closely related issue is the extent to which the programmes achieve their own targets (which in most cases seek an improvement on the initial situation). Here again the results are discouraging. They are, however, fairly consistent, with most studies showing that roughly half of the programme outcomes are below target. The results of the two more recent such studies are typical. Heller *et al.* (1988: Table 4) shows six out of twelve observations below the current account BoP target and six above; and Edwards (1989b: Table 5) shows 52 per cent on or above target for the same variable. Both studies were based on 1983–5 data. An unpublished 1991 staff study of SAF/ESAF results shows 11 outcomes better than the programme current account target, but 17 below target. The only reported target–actual comparison applied to the overall balance (reported in Reichmann, 1978) produced a similar 50–50 split. The results are similar in the various studies for the inflation rate, although with a tendency for a rather smaller proportion of targets to be met.

Various explanations are possible here. First, we should remember that quantified targets are only as good as the models which generate them, so the results just described may be influenced by ill-specified models and inaccuracies in the data fed into them. Another explanation which has sometimes been offered is that Fund staff deliberately set ambitious targets in order to influence expectations and thus help to produce an economic improvement. There may be situations in which that would be a reasonable line of action but it could scarcely provide a general reason, because to act *systematically*, or even frequently, in such a way would destroy the credibility of the exercise.

Apart from a natural tendency to over-optimism, a more likely explanation is the pressure which Fund missions are often under to reach an agreement in the face of constraints on their ability to change policies, or on the power of those policy changes to bring about improvements (Martin, 1991). In such circumstances, there is a temptation to massage the figures, to come up with a plausible-seeming set of projected outcomes in support of an agreed package of measures, even though the mission privately has doubts about the feasibility of the predicted outcomes. The survey presented on pp. 117–18 reports a specific example of this in the case of Jamaica. Similar forces are at work in heavily indebted developing countries, where IMF involvement in debt-relief operations is likely to increase pressures to agree programmes even when its missions may privately doubt their feasibility, for there may be much creditor and other pressure to conclude an agreement (this case is argued convincingly by Brown, 1992). The setting of unrealistic targets matters, however, because it contributes to the frequency of programme breakdown reported in the study of programme non-completions below, increases the probability of under-funding, and undermines the credibility of what the Fund is seeking to do.

## Comparisons with other countries

This brings us to the question whether programme countries do better than non-programme countries. This is what the with–without tests seek to tell us, although we should bear in mind the problems of achieving comparability between the two country samples, and the consequential bias in the results to exaggerate programme effects.

Here again the results are mixed but are more favourable for programme countries in the more recent tests. Thus, for both the current account and (to a lesser extent) overall BoP indicators, Khan (1990) found programme countries to have significantly better outcomes than non-programme countries. Gylfason (1987) similarly obtained a significant positive effect for the overall BoP, on 1977–9 data. Loxley's (1984) tests for *low-income* developing countries in the period 1971–82 also showed a better BoP outcome by comparison with non-programme countries, although the differences between them failed standard significance tests. The results of these and other with–without comparisons are also to the advantage of programme countries when it comes to the inflation record, with every such study showing a low inflation rate by comparison with non-programme control groups, albeit with generally low significance levels.

## Comparisons with the counterfactual

For reasons given earlier, the 'big' question remains the counterfactual: are IMF programmes associated with better BoP and inflation outcomes than would otherwise have occurred? We shall never know, of course, but, aside from in-depth country studies, 'comparison of simulations' and 'generalised evaluation' studies are the most serious attempts to get to grips with this unknowable. How do these come out? With mixed results. Khan and Knight's 1981 and 1985 simulation studies find significantly beneficial effects from policy measures of the type usually included in IMF programmes for both BoP indicators and the inflation rate, although this approach does not directly test programme effects *per se*. Doroodian (1993) arrives at similar conclusions, with strong, statistically significant, improvements in the BoP current account and weaker, non-significant, reductions in inflation.

The more recent literature has favoured the generalised evaluation approach instead. However, Goldstein and Montiel's (1986) innovatory use of this technique (on 1974–81 data) yielded no statistically significant results at all for the IMF dummy variable. Khan's 1990 study found non-significant improvements in the two BoP indicators in the twelve months following introduction of the programme and significant improvements in the following year. He found no significant change in the rate of inflation. Overall, then, the few available counterfactual tests provide only limited encouragement to the IMF, on a less than overwhelming evidential base.

## The persistence of results

One of the long-standing criticisms of the Fund's approach to BoP policy in developing country circumstances is that it is too short term, with the suggestion that any beneficial effects are ephemeral. What light does the evidence throw on this question?

Here too the results are mixed. Loxley (1984) examined with–without outcomes over one and three years. The overall balance comparison was more favourable for programme countries in the third year but the opposite was true for the current account, and in all cases his results were non-significant. However, he did find a significantly reduced inflation rate in Year 2 – a stronger result than for Year 0. We have already mentioned that Khan's (1990) generalised evaluation tests obtained rather stronger BoP results in his Year 1 by comparison with Year 0, although this was not the case with inflation. Gylfason's (1987) results indicated a weakening BoP outcome, by both measures, in Year 1 against Year 0.[14] Presumably related to this fact was his further finding that, after a substantial drop in Year 0, the expansion of domestic credit had returned to pre-programme rates by Year 1. His finding that the results of programmes which incorporated a currency devaluation were better sustained than those which did not is also pertinent. The before–after analysis on pp. 73–4 reports generally favourable results on sustainability and, overall, the available evidence tends not to bear out those who claim that programme effects are largely ephemeral.

## Programme costs

Another criticism is that the programmes impose excessive costs on the countries adopting them. We start by summarising the considerable evidence on their impact on economic growth before turning to other evidence.

The measured *impact on GDP growth* depends on the type of test. Before–after tests are almost unanimous in finding that Fund programmes are not associated with any significant change, positive or negative, in the growth rate. The only exception to this was Goldstein and Montiel's (1986) finding of a significantly negative association. Much the same result emerges from with–without comparisons: no significant differences are observable.

More adverse results emerge from other studies, however. One of the strongest is that the shortfall of actual outcomes against programme targets is the greatest in the case of GDP growth. Heller *et al.* (1988: Table 4) show only five of thirteen observations on, or better than, target; Edwards (1989b: Table 5) records an average success rate of only 28 per cent;[15] Zulu and Nsouli (1985: Table 4) for African countries record growth at or above the target rate in only five of twenty-six cases. It seems that Fund staff are especially unrealistic in their growth targeting, although we have suggested earlier that growth is not best regarded as a target variable in many programmes, in which case what these findings are recording are more in the nature of systematic forecasting biases.

The work with which Khan is associated provides grounds for believing there may, indeed, be measurable costs in terms of reduced output. First, the simulation tests undertaken with Knight (1981, 1985) predicted significant short-term reductions in growth as a result of IMF-type demand-management programmes. Second, his 1990 generalised evaluation study found significantly reduced growth rates in both Years 1 and 2 using data for 1973–88 (although not when the data were broken down into sub-periods). Goldstein and Montiel (1986) obtained no significant result on this variable, and neither did Doroodian (1993). Special interest lies in the results of a comparison undertaken by Khan and Knight (1985: Chart 1) between the predicted effects of programmes that were confined to demand-management measures and others which (like EFF and ESAF programmes) incorporated supply-side measures which would increase investment. Under the latter regime the initial loss of output is smaller, the subsequent recovery steeper and the longer-term growth trend settles down about 2 per cent per annum higher.

What now of evidence on *other programme costs*? Here there is only incomplete evidence. Loxley (1984) tested for association between

IMF programmes in low-income developing countries and changes in *saving and investment rates*, finding savings to rise in Year 0 but to have declined by Year 2, and investment rates to be lower throughout. Although none of these results were statistically significant, they are consistent with Khan's results just reported. They are also consistent with the results of evaluations of the effects of World Bank structural adjustment programmes, which are often linked to IMF programmes and which consistently show low post-programme investment levels (see Mosley *et al.*, 1991: Chapter 9; World Bank, 1989; Corbo and Rojas, 1991; Corbo and Webb, 1991).

A fairly frequent allegation is that Fund-type programmes increase poverty and/or income inequalities, but this has proved very difficult to test quantitatively. The most firmly based evidence bearing upon distributional effects is that provided by Pastor's (1987) with–without tests. His study is particularly concerned to test for correlation between programmes and changes in the share of wages in the functional distribution of income, finding a strongly significant negative relationship. Although we are not aware of other quantitative studies of this type, his result is consistent with more qualitative and anecdotal evidence that urban real wages are often adversely affected by IMF-type programmes. However, caution is needed in interpreting this as evidence of adverse net effects on poverty and inequality. Such effects will depend on the importance, and relative earnings, of the urban wage labour force, and the anatomy of poverty in the country. It is also possible that Pastor's results are stronger because he confined himself to Latin American countries. The Heller *et al.* (1988) study of the poverty effects of IMF programmes was largely inconclusive, in line with other studies,[16] although it did conclude that some programme components had aggravated the plight of certain vulnerable groups.

Although only a few studies of the poverty effects of adjustment have confined themselves to IMF programmes, there is a considerable empirical literature which looks more generally at the impact of structural adjustment programmes (SAPs), whether initiated by the Fund, the World Bank or domestic governments (see, in particular, Bourguignon and Morrisson, 1992). Unfortunately, this too yields few simple generalisations. The impact of a SAP is found to vary across poverty groups and according to the specific policies employed, which makes it hard to generalise about programme effects. There are often major differences in results as between the urban and rural poor, with the former being hardest hit (a result consistent with Pastor's). The poorest of the poor, being marginalised, are less at risk: 'the not quite so poor' are most in jeopardy – but they also stand the best chance of gaining.

One specific criticism often levelled against the Fund in this context is that its programmes result in disproportionate cuts in social services, but research on the incidence of government spending cuts shows that social services are among the more protected categories. There *have* been serious declines in health and education service provision in Africa and Latin America but most studies do not find a direct connection between this and the adoption of adjustment programmes (see Cornia and Stewart, 1990; Harris and Kusi, 1992; Hicks, 1991; Pradhan and Swaroop, 1993).

Overall, our finding of a decade ago about the distributional effects of Fund programmes still appears generally valid (*The Quest*, p. 246):

> programme effects are likely to be quite complex. Depending on the characteristics of the economy, the programme in question and the political priorities of the responsible government, the net effect could be to increase or reduce [income] concentration; there is nothing intrinsic to the logic of stabilisation that *requires* inequalities to increase. Given our findings on other variables, however, it is most likely that the majority of programmes have no statistically significant effects one way or the other, although there can be specific exceptions in either direction.

Finally, in a study not mentioned hitherto, there is quantitative evidence bearing on complaints that IMF programmes increase *political instability*. There are a number of instances where attempts to implement the provisions of a Fund programme have led to riots and other destabilising events, but are these isolated instances in politically fragile situations or are they more characteristic? Once again, there is the difficulty of distinguishing programme effects from those of the economic crises which often precede the calling-in of the Fund. After attempting to control for this and other non-programme influences, Sidell (1988) concluded from a large sample of cases that IMF programmes had not significantly increased political instability. He did not even find any correlation with episodes of collective protest, although he did suggest that governments adopting a single programme (as against repeated ones) were more likely to experience protests. One major limitation of this study, however, is that it relates only to 1969–77. A more up-to-date study is available for 1976–85, but confined to sub-Saharan Africa (Moore and Scarritt, 1990). This similarly (albeit reluctantly) concludes that Fund programmes have no significant impact on the nature of African governments.

Overall, then, there is little evidence that Fund programmes impose heavy costs on the economies affected – which is not perhaps surprising since much of the other evidence already surveyed shows how easy it is to exaggerate programme impacts on other variables. Before we turn for explanations of this relative nugacity, however, there are two other points worth bringing out.

First, it can be hypothesised that countries at a lower level of economic development find stabilisation and structural adjustment particularly difficult because of structural weaknesses and inflexibilities in their economies (*The Quest*, Chapter 8; Killick, 1995: Chapter 12). The most direct evidence on this is that provided by Loxley (1984: Table 5.1) who compared results for a sample of least developed countries with those for a sample used by Donovan (1982), which included a substantial number of middle-income developing countries. For both BoP indicators he found his results were worse than for Donovan's sample, although the results were more mixed for other variables. We might also note the particularly adverse evidence on programme effectiveness provided in Zulu and Nsouli's (1985) before–after and target–actual tests mentioned earlier, which may well have been because their tests were restricted to African (and hence mainly least developed) countries.[17]

Secondly, we may also ask whether there is any discernible trend in programme impact over the last two or three decades. It seems not, although we should bear in mind that only one or two of the studies reported have made much use of data for programmes in the 1980s. The best evidence is provided by Khan (1990: Table 6). He disaggregates for 1973–9 and 1980–8, finding a weaker current account but much stronger overall BoP effect in the 1980s. However, only the improvement in current account was significant at the 95 per cent level for the 1980s (although the overall BoP was nearly so), whereas both the current account result and an association with higher inflation were significant for the 1970s. The only other study which encompasses the 1980s is Edwards's (1989b) target–actual comparisons, but we have already noted that his results are in line with those of other target–actual comparisons relating to earlier periods. Once again, we have to report no strong conclusion. A further test of this is reported on p. 63.

Given the apparently quite considerable extent to which Fund programmes fail to have their intended results, we would like to have evidence on the reasons for the shortfalls. Unfortunately, most of the literature under review does not address this issue much, although it provides some evidence.

## Sources of difficulty

### *The initial situation and exogenous shocks*

It is self-evident that the severity of the initial situation will have a strong influence on the likelihood of programme 'success'. There is similarly little dispute that large disturbances in the global economy have knocked many programmes off course, particularly when borrowing countries experienced major changes in their terms of trade. Moreover, the effect may be asymmetrical: when there is a large sudden terms-of-trade deterioration programmes 'fail' because the original policies cannot cope with the worsened situation. But when the terms of trade suddenly improve programmes are also apt to 'fail', or rather to be abandoned, because reduced BoP pressures allow governments to feel that they can relax policies (for an illustration of this process, relating to Kenya in the 1970s, see Killick, 1984b: Chapter 5; see also Killick and Mwega, 1993).

Although the studies examined here provide only limited evidence on the importance of the initial situation and of exogenous shocks, Khan (1990: Table 4) does broach these subjects. For each of his target variables he finds that easily the most significant 'explanation' is provided by the previous year's value of the same variable, which places a large implicit weight on the initial situation. He also finds changes in the terms of trade to have highly significant influences on the BoP and GDP growth. Goldstein and Montiel's analysis (1986: 337–8) is less explicit, but shows how pre-programme conditions can make a major difference to the outcome of before–after and with–without comparisons.

Doroodian (1993: 858) similarly obtains significant values for the current account influence of the terms of trade.

### *Impact on policies*

Another possible reason for the low impact of programmes is that they do not have sufficient influence on policies to make much difference to BoP or other outcomes. The first suggestion that this might be the case was given by Reichmann and Stillson's 1978 before–after study. They found there was no significant change in the rate of expansion of total domestic credit (a key policy instrument in IMF programmes) in two-thirds of the cases studied; the same was true for credit to the public sector in half of the programmes, although there was a significant deceleration in two-fifths of the cases. *The Quest* (Table 7.4) similarly found no significant change in the growth of domestic credit on a before–after basis.

Also on data relating to the 1970s, Gylfason (1987) found a highly significant initial reduction in credit expansion, but that by Year 1 this was nearly back to the pre-programme rate, while monetary expansion was a good deal higher (32 per cent per annum against 25 per cent – see his Table 2). However, in their study of programmes in Africa, Zulu and Nsouli (1985: Table 4) found that for total domestic credit about half of their observations were on target or better, on both with–without and before–after comparisons; and Stuart (1991: Table 2) provides evidence showing that credit targets were met in 27 out of 44 stand-by programmes in the period 1985–8.

In both the last-mentioned studies the outcome with attempts to reduce budget deficits was weaker. Zulu and Nsouli found only eight out of 27 budgetary out-turns were on target or better, and on a before–after basis only 12 out of 29 budget deficits were lower than in the pre-programme year. Stuart's data show fiscal targets not to have been met in nearly half (20 out of 44) of stand-by programmes. We should also mention information provided for programmes in the early 1980s by Heller *et al.* (1988: Table 5), which shows six out of eleven observations for credit to the public sector at above-target levels. True, their statistics on credit to the private sector reveal only three out of eight above target, but that is an ambiguous result because it may well have been a programme objective to raise the proportion of credit going to the private sector.

By far the most substantial evidence on fiscal effects, however, is provided in Nashashibi *et al.*'s 1992 study of the fiscal dimensions of SAF/ESAF programmes in 23 low-income countries. This finds only a rather slight tendency for programmes to be associated with improved fiscal balances, on a before–after basis, with 13 out of 23 countries obtaining reduced (or not increased) deficits, and with a difference in the before and after means apparently falling far short of standard statistical tests for significance. Finally, a study by Cashel-Cordo and Craig (1990) studies the policy leverage of different forms of aid, where IMF credits are included as a form of aid. They found Fund credits to be associated with reductions in *both* government expenditures and revenues, with a larger relative decline in revenues. This result was statistically significant for stand-by but not for EFF credits. They suggest that Fund credits have substantial leverage in 'buying' quite large fiscal effects, but observe that the Fund is better at shrinking the total size of the public sector than at reducing the budget deficit.

### *Programme execution*

Overall, the above suggests rather strongly that the Fund is unable to change policies to the extent that it would like to – and to the extent that would be necessary for programme success. One possible explanation of this is incomplete programme implementation. Fund staff members often cite 'lack of political will' as a common reason for programme breakdown, and the study of programme non-completions on pp. 56–66 demonstrates frequent incomplete implementation of stand-by provisions. The most recent published evidence is provided by Edwards, who examines the compliance record for 34 programmes approved in 1983 in respect of the government budget deficit, total domestic credit and domestic credit to government, and finds that, overall, less than half (45 per cent) of all such conditions were observed (1989b: Table 4). Observance of programme requirements was particularly weak for the government budget; strongest for credit to government. A similar story is told by earlier studies (see *The Quest*, 1984a: Table 7.3).

An unpublished 1991 staff review of experiences with SAF and ESAF programmes throws some additional light. Slightly over half of all benchmark criteria (i.e. the most binding aspects of programme provisions) were observed on schedule, or two-thirds within a few months thereafter. However, the pattern of implementation is interesting, with full compliance with conditions on agricultural producer pricing and marketing; quite high implementation of financial sector reforms; only 50 per cent implementation of fiscal provisions; and particularly slow progress on public enterprise reforms.

This evidence should be evaluated against fairly strong expectations that, *were they implemented*, the types of policy change commonly incorporated in Fund programmes would produce substantial beneficial economic results. This is the main thrust of the comparison-of-simulation exercises reported above. Consistent with this, the limited publicly available evidence suggests fairly strong correlations between implementation and programme results. As can be seen in Table 3.2, Stuart's analysis of stand-by outcomes shows markedly stronger results for economic growth and the BoP (but not for inflation), according to implementation of fiscal and monetary provisions.

*The Quest* (pp. 257–9) also found a general, if imperfect, association between implementation and programme results, although Connors (1979) obtained a weaker result. He divided his programme countries into compliers and non-compliers and tested

*Table 3.2* Programme implementation and economic performance

| | *Full implementation* | *Low implementation* |
|---|---|---|
| GDP growth (% p.a.) | | |
| Pre-programme | 1.8 | 2.9 |
| Programme period | 4.4 | 1.0 |
| Inflation rate (% p.a.) | | |
| Pre-programme | 20 | 93 |
| Programme period | 19 | 78 |
| BoP current a/c (% GDP) | | |
| Pre-programme | −5.3 | −7.8 |
| Programme period | −2.4 | −5.1 |
| Overall BoP (% GDP) | | |
| Pre-programme | 0.5 | −2.7 |
| Programme period | 1.6 | −3.5 |

*Source*: Stuart, 1991, Table 3.
*Note*: 'Implementation' here refers to whether or not programme fiscal and credit targets were met.

for association between programme results and compliance. Although most of his signs were 'right', only one, relating to inflation, was statistically significant.

### *Catalytic effects*[18]

It is a long-standing claim of the IMF that its programmes have a catalytic effect, with the 'seal of approval' signified by a programme agreement triggering additional public and private sector capital inflows. A more recent variant on this is that agreement with the Fund is often a formal prerequisite for debt reschedulings or reductions through the Paris and London Clubs. That programmes should have such catalytic effects is important for their success, because the resources of the IMF itself have been allowed to decline, relative to trade values, to the point that the credits it can make directly available in support of a stabilisation programme are limited, relative to a country's total financing needs.

A final question we can explore, therefore, is whether programme effectiveness is undermined by an insufficient response by other sources of capital inflow, causing programmes to be under-funded. Forming a judgement on this is by no means easy because the capital account includes a variety of capital flows and we would not expect them all to respond equally to agreement on a Fund programme. There may also be important financial flows 'above the line', most notably interest payments on external debt. Indeed, there may be 'perverse' reactions. The clearest example of this was in some heavily indebted countries during the 1980s where IMF credits were, in effect, used to service commercial bank debts.[19] It is also possible that potential creditors or investors could take a Fund programme as a signal of previously undetected financial difficulties and defer new commitments.

Past evidence has suggested that we should not in general expect major net catalytic effects. A survey of the evidence relating to the 1970s arrived at this conclusion (*The Quest*, pp. 235–6). Comparison of the results reported above for programme effects on the current and overall BoP produces a similarly cautionary conclusion. Were there a strong catalytic effect, we would expect the reduction in the overall BoP deficit (or increase in surplus) to be substantially larger than for the current account but, although one or two studies do report such an outcome, no such general result emerges clearly from the empirical studies reported above (e.g. see Table 3.2).

Moreover, our own results reported on p. 68 indicate the current account improvement to be rather *larger* than for the overall balance, suggesting a perverse movement of capital.

Although neither of them directly tests the results of IMF programmes *per se*, both Hajivassiliou (1987) and Faini *et al.* (1991) produce results casting further doubt on catalytic effects. The former found for 1970–82 a *negative* relationship between IMF support and new private sector lending; the latter found a significantly negative correlation between combined Fund and World Bank lending and net private credit in 1982–6.[20] Recent results of a detailed analysis of this topic by Rowlands (1994) also reveal little net impact on private creditors. However, he does also find that lending from *official* sources was strongly responsive to stand-by and EFF programmes, especially for lower-income countries post-1981, so that, on balance, total net capital inflows did increase significantly. There was insufficient evidence on SAFs/ESAFs.

The evidence is thus mixed, but leaves the probability that programme effectiveness in specific countries (perhaps middle-income ones in particular, for whom private capital markets are the chief source of new credit) is undermined by continuing shortages of foreign exchange and the depressing effects these are liable to have on export performance, government revenues and economic activity.

## A STUDY OF PROGRAMME NON-COMPLETIONS

We now turn from reviewing the evidence in already published studies to report the results of research undertaken at the Overseas Development Institute, starting with an analysis of the substantial proportion of Fund programmes which are abandoned before the end of their intended life, and continuing, in the next section, with a before–after analysis of experiences in the 1980s.

### Explanation and justification

With patience and ingenuity it is possible to assemble from publicly available sources a record of the extent to which IMF programmes are drawn down as intended (although this is certainly one of the areas in which the operations of the Fund should be made more transparent).[21] The basic premise of this analysis is that the discontinuance of a programme before the end of its intended life is a useful indicator of performance under that programme. Why this should be so requires explanation.

The Letters of Intent which are formally addressed by governments to the Fund setting out agreed policy measures identify a number of performance criteria, as discussed in Chapter 2. A proportion of the IMF credit associated with the programme will be payable at the time of its approval, with the balance payable in instalments or tranches. Unless the terms of the agreement are subsequently changed by the granting of a waiver by the Fund, access to outstanding tranches will be conditional on observance of the performance criteria. Thus if, say, credit to government goes above the specified ceiling, access to any outstanding tranches of the credit will be suspended (unless a waiver is granted) until the figure is brought back within the ceiling. A high proportion of programmes suspended in this way are subsequently abandoned, although they may be followed quickly by a new agreement.

Interruptions in IMF credit disbursements thus tell us that the government has been unwilling or unable to conform to the agreed terms and that key economic magnitudes have exceeded (or gone below in the case of minima) the levels deemed by Fund staff to be consistent with programme objectives. Since most performance criteria refer to policy variables, this also gives us information on the impact of programmes on fiscal, monetary and other policies. However, the results of this test should be used with caution because of the opaque way in which the Fund publishes the information and because it sometimes happens that programmes are allowed to lapse by mutual agreement, chiefly owing to unforeseen developments which render the targets or other aspects of the programme unattainable. In these cases programme breakdown does not imply delinquency. On the other hand, while observance of performance criteria is a necessary condition of programmes achieving their objectives, it is not a sufficient condition, so for this reason too our indicator can only be a partial one. Programmes may be mis-specified; policy changes may have insufficient force or be undermined by the adverse effects of other policies; external conditions may deteriorate: for these and other reasons even programmes which successfully meet their performance criteria may often fail to achieve their objectives.

A further factor to bear in mind is that most programmes have in the past been in the form of 'stand-bys' (although that has changed recently with the growth in the number of EFF and

ESAF programmes) and that the provisions in such programmes typically extend for 12–18 months, at the end of which time all tranches are expected to have been released by the Fund. It ought not, for such programmes, to be an excessively harsh criterion to test whether implementation remains within agreed parameters for such a relatively brief period. The position is different in the case of EFF and ESAF programmes, which normally have an intended life of three years, and for which we might expect higher rates of non-completion.[22]

For the purposes of the present study an uncompleted programme is defined as one in which 20 per cent or more of the total value of the credit remains undrawn. What this means in most cases is that the programme is discontinued at least before the final credit tranche is released. Of course, in many of the cases of non-completion the undrawn balance was considerably greater than 20 per cent.

Although the test just described is a limited one, it has the advantage that, once the basic information has been assembled, it is a straightforward matter to look for regularities in the pattern of non-completions. The following analysis is based on information on all the 353 stand-by, EFF and SAF programmes approved in the period April 1979–April 1993. However, as at April 1993, 35 of these were still in the process of implementation and thus had to be excluded from our calculations. We also excluded 13 SAF programmes which had been converted into ESAFs which themselves were still current. It did not seem appropriate to treat such conversions as breakdowns. We were thus left with 305 programmes for analysis.

We were, happily, in a position to form an independent judgement on the reliability of our 20 per cent test as an indicator of programme performance. At the time this work was first undertaken we were also engaged in the survey of country-specific case studies reported on pp. 86–119.[23] Because these were based on in-depth research by authors knowledgeable about the countries they were analysing, these provide better, and independent, information on programme performance. This survey took in 17 countries encompassing a total of 48 programmes in the 1980s. We were thus able to compare the information provided in the country studies with the results of our much simpler test based on programme completion.

The outcomes matched almost perfectly. Although the work was undertaken separately and the comparison was not made until the rest of the work had been completed, an identical match was obtained on the 29 stand-bys common to both exercises, while there was only a small difference on EFFs, for which the 20 per cent test indicated that 13 out of 16 programmes failed during the course of their intended life while the (more accurate) country studies indicated a 15 out of 16 score. The highly satisfactory outcome of this comparison confirms the accuracy of our information on programme breakdowns and supports the usefulness of this as an indicator of programme performance, adding weight to the significance of the results reported below.

## The results

The results are summarised in Table 3.3. Just over half (53 per cent) of the 305 programmes were uncompleted in the sense defined. However, the results differed according to programme type, with the best results (50 per cent non-completions) for stand-bys and by far the worst for EFFs (85 per cent). The 38 per cent result for SAFs should probably be discounted because of the small number of observations.

The specially poor result for the EFF programmes can be related to the controversies which surrounded such programmes early in the 1980s. A considerable number of these were signed in the late 1970s and 1980-81 but use of this facility was then almost abandoned until the early 1990s. IMF staff defended this change of policy by asserting that performance under these programmes had been poor, while others (including the present writer) were critical of the change, disputing that the EFF's record was specially weak and deploring the retreat from a facility which permitted the Fund at least to take a medium-term view and to include more supply-side measures in the programmes (see *The Quest*, pp. 211–12 and 247–50).

In retrospect it is clear that the staff were right in their judgements about the EFF. By our non-completion test, and even though we would expect a higher slippage rate on three-year as against one-year programmes, its record seems little short of disastrous (a judgement supported in the survey of country experiences with IMF programmes). It is therefore rather surprising that at the beginning of the 1990s the Fund again began to approve EFF programmes. It will be interesting to see whether they fare better than a decade earlier.

*Table 3.3* Analysis of IMF programme completion, 30 April 1979 to 30 April 1993[a]

| | *Uncompleted (%)* |
|---|---|
| 1. *All programmes* (305) | 53 |
| of which | |
| Stand-bys (251) | 50 |
| EFFs (33) | 85 |
| SAFs (21) | 38 |
| 2. *By period* | |
| 1979/80–1982/83 (91) | 44 |
| 1983/84–1985/86 (68) | 41 |
| 1986/87–1989/90 (54) | 56 |
| 1990/91–1992/93 (28) | 61 |
| 3. *By region*[b] | |
| Sub-Saharan Africa (115) | 50 |
| Western Hemisphere (64) | 53 |
| Asia (30) | 47 |
| 4. *By income category*[b,c] | |
| Low-income (182) | 47 |
| Lower-middle (220) | 53 |
| Upper-middle (73) | 53 |
| 5. *By debt status*[b,d] | |
| Severely indebted (132) | |
| Low-income (82) | 52 |
| Middle-income (50) | 60 |
| Moderately indebted (70) | |
| Low-income (25) | 36 |
| Middle-income (45) | 43 |
| Others (48) | 46 |
| 6. *By dominant export*[b,e] | |
| Primary products (156) | 56 |
| Fuel (21) | 67 |
| Non-fuel minerals (40) | 40 |
| Agricultural (95) | 60 |
| Manufactures (30) | 43 |
| Diversified exports (31) | 39 |

*Notes*: Numbers in parentheses are numbers of programmes.

[a] Analysis excludes all programmes still current at 30 April 1993 as well as all SAF programmes which were converted to ESAF programmes.

[b] Analysis relates to stand-by programmes only.

[c] Reported results are the means of results obtained by using 1980 and 1988 country income classifications in World Bank, *World Development Reports* of 1982 and 1990 respectively.

[d] Classification according to World Bank *World Debt Tables* 1993–94, Vol 1.

[e] Classification according to IMF, *World Economic Outlook*, October 1989.

Given the preponderance of stand-bys in the period (251 of our 305 programmes) and the brevity of their intended disbursement period, their 50 per cent non-completion rate is the outstanding fact to emerge from the analysis. The IMF can scarcely be satisfied with a situation in which half such programmes break down so quickly. Moreover, further analysis revealed that 18 per cent of all stand-bys (or 36 per cent of the discontinued ones) broke down almost immediately, with little or no utilisation of the credit beyond the portion that was payable on approval of the programme.[24]

Items 2 to 6 of Table 3.3 set out further analyses of our data which, in some respects, enable us to go further than the literature surveyed in the previous section. Analysis of these items is confined to stand-bys to avoid the distortions that would be introduced by including EFFs. First, we can ask about completion *trends over time*. From this it emerges that non-completion was a growing problem from the mid-1980s. Three-fifths (61 per cent) of all programmes were uncompleted in the latest period shown, with the difference between this and the 1983/84–1985/86 average of 41 per cent significant at the 95 per cent level. This shows the difficulties the Fund continues to have in attempting to improve on the poor past record on programme compliance. There is no evidence of an improving trend here; quite the contrary.

We next examined whether programme completion was related to *regional factors*. The hypothesis was that programmes in African and Latin American countries would have less success than those in Asia. In the African cases this could be predicted because of generally deteriorating terms of trade and structurally weak economies, and in Latin America because of the negative effects in many of them of a large debt overhang. Table 3.3 suggests that there may be something in this, with Asian countries revealing the lowest non-completion rates, but the differences are not large (or significant).

On the hypothesis that the least developed countries would have the greatest difficulties in executing Fund programmes successfully (for reasons sketched earlier), we examined whether programme completion rates were a function of country *per capita income* levels, using the World Bank's system of country income classification. Table 3.3 shows that this hypothesis was conclusively rejected, with no significant differences in completion rates between the three income categories. A possible reason for this result is that in the 1980s and early 1990s any relationship between programme viability and level of development is liable to have been obscured by the special problems of heavily indebted countries, a high proportion of which are classified in the middle-income categories.

Countries' *debt positions* yield an apparently stronger explanation, however, as can be seen from item 5 of Table 3.3, even though the results only pass a 90 per cent significance test. While over half of programmes entered into by severely indebted countries broke down, this was true of only about two-fifths of moderately indebted countries with small external debts. We can conjecture that this result is attributable to the severe strains which the governments of heavily indebted countries are already under at the inception of a programme. In particular, severe external indebtedness means that large amounts of public revenues must be devoted to the servicing of the debt, making it all the more difficult to reduce fiscal deficits in the way that is normally required in IMF programmes. Moreover, the use of currency devaluations is less appropriate in such situations and where the debtor country's amortisation payments have made it a *de facto* net exporter of capital to the rest of the world (see Reisen and van Trotsenburg, 1988), which reduces the number of effective policy instruments available to make a programme work. Finally, the economic hardships already imposed on the populace by the debt burden are likely to make strict adherence to the further austerities of IMF-sponsored programmes very difficult politically.

One other point of note is the heavy concentration of stand-bys on severely indebted countries shown in the table: 132 out of the 250 classified by debtor status. Since many of these programmes were necessary preconditions for debt rescheduling agreements in the Paris and, to a lesser extent, London Clubs, this result suggests that in many cases the debt relief was not enough to be consistent with programme success (for a detailed treatment with reference to African countries, see Martin, 1991). These results tie in with findings of weak catalytic effects with respect to private capital, reported earlier (see pp. 57–8).

Closely related to the above is the *adequacy of the IMF's own credits*. It is sometimes suggested that these are small in relation to need, especially when one takes into account return flows to the Fund in respect of earlier credits. We therefore undertook an analysis of the magnitude of the credits drawn down in our sample of 38 programmes in 1979–89 and related these to BoP magnitudes. The results are summarised in Table 3.4.[25]

*Table 3.4* Indicators of the adequacy of Fund credits, 1979–89 (%)

| | *Annualised mean use of IMF credits as % of:* | | | | |
|---|---|---|---|---|---|
| | *GDP* | *Base-period values of:* | | | *Return flows during programme* |
| | | *Current a/c deficit* | *Overall BoP deficit* | *Imports* | |
| | *(1)* | *(2)* | *(3)* | *(4)* | *(5)* |
| All programmes | 1.8 | 29.3 | 22.6 | 9.9 | 1,205 |
| Completed programmes | 2.6 | 48.3[a] | 73.4 | 16.9 | 1,105 |
| Uncompleted programmes | 1.3 | 18.6[a] | −6.2[b] | 6.0 | 1,261 |
| 1979–82 programmes | 2.1 | 25.7 | 0.2 | 7.9 | 862 |
| 1983–85 programmes | 1.4 | 33.4 | 47.5 | 12.2 | 1,588 |

*Notes*:
[a] The difference between this pair of average values is significant at the 90 per cent level.
[b] The negative sign indicates that there was an initial overall surplus.

Of course, what constitutes 'adequacy' is a matter of judgement and will vary from case to case. None the less, if we take all 38 programmes together, it can be seen that credits utilised were equivalent to three-tenths of the pre-existing deficit on current account, nearly a quarter of the overall BoP and nearly 10 per cent of total imports. They were therefore far from insignificant. Moreover, the credits utilised exceeded by a factor of 12 the return flow of amortisation ('repurchase') payments in respect of previous programmes (column (5)). The Fund was thus providing quite substantial *net* assistance. This needs to be qualified in three respects, however. First, we were not able to include data for interest payments in our 'return flow' figures, which means that net resource transfers were somewhat smaller than shown in Table 3.4. Second, the adequacy indicators were substantially lower during 1979–82 than in the following years. Third, but not shown in the table, the net flow was substantially smaller for countries which had just successfully completed an earlier programme. For them the net flow was only 231 per cent of repurchases.

Of more direct relevance to the present discussion, however, are the differences between the positions of the countries which did and did not complete their programmes, for it can be seen that most of the indicators of adequacy are substantially smaller for the non-completers, even though these annualised data are adjusted for the effects of programme suspensions and cancellations. In other words, relative to imports and BoP balances, the non-completers received substantially smaller credits than the others – a difference that was, however, only significant at the 90 per cent level. Although it can be no more than suggestive, this result is consistent with the hypothesis that inadequate supporting finance is a reason for programme breakdown.

We further looked for any association between frequency of programme breakdown and *type of export* (Table 3.3, item 6). The hypothesis here was that exporters of manufactures would have above-average completion rates because, being sellers on generally buoyant world markets, they would find it easier to boost export earnings by comparison with exporters of (non-fuel) primary products. This hypothesis receives only partial support. Exporters of manufactures turn out to have only moderately below-average breakdown rates, while exporters of non-fuel mineral products have a slightly better record. Programme breakdowns among agricultural exporters do conform to our expectations, however, with a 43 per cent breakdown rate (statistically significant at the 95 per cent level). Interestingly, the lowest rate belonged to countries classified (by the IMF) as having diversified export bases.

There is a further use to which we can put the information assembled for the above analysis, pertaining to the *sustainability of programmes*. IMF assistance is intended to provide *temporary* support to governments seeking to restore viability to their countries' BoP. It was not expected to be providing frequent credits over a sustained period. However, from our inventory of all programmes in effect during the period April 1979 to April 1993 (excluding SAFs converted to ESAFs but including ESAFs), it is clear that repeated assistance is frequent. In fact, over these 14 years no less than 23 countries had six or more programmes approved by the Fund, encompassing 162 programmes, or 44 per cent of the total for the period. Such programmes were often approved back-to-back or with only brief intervening periods. The record was nine programmes in the 14 years, a distinction shared by Madagascar, Senegal and Togo (all from the African region, be it noted, and two from the Franc Zone countries). While the Fund should be given credit for its persistence, its founding fathers cannot be happy that its programmes have been unable to bring sustained improvements to the payments positions of these countries.

## A BEFORE–AFTER ANALYSIS FROM 1980

### Nature of the study

We now turn to offer additional information concerning programmes adopted in the 1980s. In the course of conducting the research on which the next section is based, statistical information was collected on macroeconomic variables for each of the countries studied and in this section we utilise this information to present a quantitative analysis, based on before–after tests. The limitations of such tests have already been described and should be borne in mind.

What follows has some advantages over most of the comparable literature. It is somewhat more up to date, being based on programmes begun and completed (or abandoned) in the period 1979–85. It examines a wider range of BoP variables so as to facilitate a more rounded view of programme effects. It includes an analysis of changes in domestic absorption and its chief components, on the grounds that a strengthening of the BoP is likely to require a reduction in total absorption relative to GDP and in order to examine further where cuts in absorption make themselves most felt. It takes us further into the neglected area of programme effects on key policy variables. It studies the effects of programmes over a longer period (up to four years from the start of the programme) in order the better to test whether programme effects were sustained. It tests for differences in results between programmes that were, or were not, completed. Finally, it tests whether the existence of another IMF programme immediately prior to that under examination made any difference to outcomes.

### Results

The results are summarised in Table 3.5. The two right-hand columns of this differentiate the Year 1 + 2 results according to whether the programmes in question were completed or not, using the same 80 per cent draw-down cut-off as that used in the previous section.

#### *BoP effects*

Improvements to the BoP were among the strongest of our results, with appreciable and statistically significant improvements in both

*Table 3.5* Results of before–after tests on 16 developing countries with IMF programmes commenced 1979–85*

| *Variable* | *Base value* (Yr −1, −2) | *Differences from base value* Yr 0 | Yrs 1, 2 | Yr 3 | Completed programmes (Yrs 1, 2) | Uncompleted programmes (Yrs 1, 2) |
|---|---|---|---|---|---|---|
| *BoP indicators* | | | | | | |
| 1 Overall BoP (as % GDP, change in reserves) | −1.2 | +0.4[a] | +1.2[a,b] | +1.9[a,b] | +0.7[a] | +1.5[a,b] |
| 2 Current a/c (as % GDP, exc. official transfers) | −11.2 | +1.1 | +3.0[a,b] | +3.8[a,b] | +4.4[c] | +2.2[a,c] |
| 3 Official transfers (as % GDP) | 2.8 | −0.2 | +0.0 | +0.0 | −1.4 | +0.8 |
| 4 Capital account | | | | | | |
| (a) Direct foreign investment (as % GDP) | 0.4 | +0.0 | +0.1 | +0.3[c] | −0.0 | +0.2 |
| (b) Net long-term loans (as % GDP) | 5.5 | −0.0 | −1.2[a] | −2.0[a,d] | −1.1[a] | −1.3[a] |
| 5 Terms of Trade index (% change)† | | −3.8[a,d] | −6.5[a,d] | −8.2[a,d] | −2.5 | −8.8[d] |
| 6 Import volume index (% change) | | −4.6[c] | −3.1 | −4.1 | +0.2 | −5.0 |
| 7 Export volume index (% change) | | +3.5[a] | +9.6[a,b] | +11.7[a,b] | +8.7 | +10.2[a,b] |
| *Inflation and growth* | | | | | | |
| 8 Increase in consumer prices (% p.a.)‡ | 24.8 | +1.7 | −1.1[d] | −0.4 | −12.7[d] | +5.6 |
| 9 GDP growth (% p.a., at constant prices) | 2.1 | −0.0 | +0.6 | +1.2 | +1.8 | −0.0 |
| *Domestic absorption* | | | | | | |
| 10 Total absorption (as % GDP) | 107.5 | −2.5[a,b] | −3.4[a,b] | −3.3[a,b] | −1.2 | −4.6[a,b] |
| 11 Private consumption (as % GDP) | 71.6 | −0.3 | −0.4 | +0.2 | +1.5 | −1.5 |
| 12 Government consumption (as % GDP) | 13.7 | −0.4 | −0.3 | −0.4 | −1.0[b] | +0.1 |
| 13 Fixed investment (as % GDP) | 21.2 | −1.9[a,b] | −3.1[a,b] | −3.8[a,b] | −3.4 | −2.9[a,b] |

| | | | | | | |
|---|---|---|---|---|---|---|
| *Policy variables* | | | | | | |
| 14 Real effective exchange-rate index (% change) | | −7.1[a,c] | −11.0[a,b] | −15.0[a,b] | −18.1[c] | −7.0[a] |
| 15 Total domestic credit | | | | | | |
| (a) growth rate (% p.a.) | 25.8 | −3.4 | −3.3[a] | −4.2[a,c] | −3.3 | −3.2 |
| (b) as % GDP | 42.4 | −0.4 | −0.7 | −1.8 | +2.0 | −2.4 |
| 16 Private sector credit | | | | | | |
| (a) growth rate (% p.a.) | 23.1 | +0.4 | +0.8 | +0.8 | −3.6[a] | +3.2 |
| (b) as % GDP | 18.9 | −0.8 | −0.8 | −0.2 | +0.8 | −1.6 |
| 17 Credit to central government (as % GDP)§ | 18.2 | +0.2 | +0.0 | −1.5 | +1.0[a] | −0.6 |
| 18 Central Government budget deficit (as % GDP) | −6.7 | +0.2 | +1.4[b] | +1.6[b] | – | – |

Notes:
[a] Significant proportion of programmes with either a positive or negative change, as compared to a hypothesised equal proportion.
[b] Significantly different from zero, under a one-tailed t-test, at the 99 per cent confidence level.
[c] Significantly different from zero, under a one-tailed t-test, at the 95 per cent confidence level.
[d] Significantly different from zero, under a two-tailed t-test, at the 95 per cent confidence level. For line 8, significance tests are based on log-linear data.

* Notes on the data and details of the various tests are available on request.
† Completed and uncompleted programme values are significantly different from each other at the 90 per cent confidence level.
‡ As for †, but at the 95 per cent confidence level.
§ Year-to-year changes in credit to government were too variable for it to be meaningful to calculate changes in growth rates.

the overall and current account balances. The improvements were relatively slight for the programme year (Year 0) but were strong for the following three years. A quantitatively larger reduction in the current account deficit was achieved in countries which completed their programmes but, curiously, the opposite was true with respect to the overall BoP.

We should note, moreover, that the BoP results were obtained in the face of rather strongly adverse movements in the commodity terms of trade (line 4), which by Year 3 had, on average, deteriorated by a full 8 per cent. Observe here that terms-of-trade experiences under the uncompleted programmes were substantially worse than with completed programmes, which strongly suggests that worsening trade conditions were a source of programme breakdown.

One possibility is that the current account results were secured simply on the basis of an import squeeze (in which case we would have to be careful about calling this an improvement) and line 6 of the table provides information on this. The statistics there show that there was a significant tendency for import volumes to be cut in the programme year, a tendency (not shown in the table) concentrated in the most recent years.[26] The impact was not large, however, nor was the result statistically strong except in the programme year. A possible explanation for this is that the crisis conditions in which many programmes were introduced resulted in reduced import volumes in the year or two immediately preceding the programme, because of reduced creditworthiness, depleted reserves, growing prior claims of debt-servicing obligations, etc. We therefore examined the behaviour of imports in the years immediately preceding programme adoption. This revealed that there was indeed a modest tendency for imports to be reduced in the pre-programme year, but this was not statistically significant. All in all, the worst fears that IMF programmes lead to sustained import strangulation were not borne out.

More positively, we found rather strong evidence that programmes were associated with improved export performance, as indicated by growth in export volumes (line 1 of the table). Given normal supply response lags, it is not to be expected that there would be any large immediate export response, but we found consistent trends (significant at the 99 per cent level or above) from Years 1 and 2 through to Year 3, particularly among those countries which had previously completed programmes.

We have earlier examined the claim that a programme with the IMF has a catalytic effect on net capital inflows. Items 3 and 4 of Table 3.5 provide further evidence on this (treating official transfers as capital transactions). As can be seen, the changes recorded are quantitatively quite small, although several are statistically significant.[27] *The average net effect is a reduced capital inflow.* One possibly important explanation of this finding is the much smaller initial deficit on the overall account – only 1.2 per cent of GDP. However, a more important explanatory factor would seem to be the increase in net repayments of long-term loans – i.e. a substantial part of the improvement in the current account was used to finance the repayment of past loans and was not rewarded by increases in disbursements.

This result is particularly surprising, given the frequent linkage of IMF programmes to debt rescheduling agreements (which would reduce the level of principal repayments). However, the practical effect of this linkage on actual financial flows was limited by the fact that debt-distressed countries were not servicing their debts anyway. In some cases, Fund credits were used to repay other creditors. We should bear in mind, however, that we are not able to make comparison with non-programme countries, which may have experienced even higher rates of net loan repayment,[28] and Rowlands's (1994) contrary evidence, reported earlier.

### Inflation and growth

Weak results are also revealed with respect to inflation (Table 3.5, line 8), with only the smallest reductions on pre-programme levels. Over 40 per cent of the programmes were associated with an *increase* in the inflation rate – even in the longer term.

More interesting are the longer-term results for real GDP growth, as they suggest that after three or four years programmes may be associated with *increases* in growth rates. However, the results reported here offer only weak evidence – the increase after four years is only significant at 90 per cent confidence levels. There is also some weak evidence that completed programmes are associated with longer-term improvements in GDP growth rates (statistically significant in Year 3 at the 90 per cent confidence level). These results are surprising, particularly in view of the reduced levels of fixed investment reported below, and imply that the growth must have come through improved efficiency of resource utilisation.

## Domestic absorption

In the absence of any strong catalytic effect, absorption has to be reduced relative to GDP if the BoP is to be strengthened. That programmes are associated with such a reduction is among our strongest results (line 10), with substantial and significant changes over the entire post-programme period. If cause and effect are at work here, this must be rated a programme success; but note the odd result that the largest reduction is in respect of uncompleted programmes, with only a non-significant fall in the case of completed programmes. This could be because initial domestic absorption as a proportion of GDP was on average 4 percentage points higher in uncompleted programme countries compared with countries with completed programmes, which reinforces the view that programme success is related to the severity of the initial situation.

But on which component of absorption do the reductions chiefly fall? The brunt falls on fixed investment, which declines substantially and significantly over the whole period. Overall, the programmes appear unable to exert any appreciable squeeze on private or public consumption, although there is a shift in its composition in favour of the private consumer in the case of completed programmes. The adverse impact on investment is consistent with results reported in the discussion of existing literature (see p. 50) and in evaluations of World Bank programmes. Where programmes have the effect of bringing excess capacity into utilisation, temporary reductions in investment can be consistent with continued economic growth, but if investment remains depressed this becomes a serious source of difficulty for policy-makers. The results in Table 3.5 suggest that investment not only remains depressed but actually continues to decline.

## Impact on policy variables

It was also suggested earlier that their relative lack of impact on key policy instruments provided an explanation of why programmes often do not meet their objectives. Items 14–18 in Table 3.5 provide evidence on this. The strongest result is with respect to the real exchange rate, which is shown as being depreciated by an average of 7 per cent in the programme year, a depreciation which is not merely sustained but deepened during the following three years, and which is particularly large for programmes which run their intended term. This no doubt helps explain the substantial improvement in export performance reported earlier.

These results, moreover, have generally high levels of significance. Although we cannot make any direct comparison with non-programme countries in this period, it is worth noting that our average rate of −11 per cent per annum over the four post-programme years is significantly larger than the average real devaluation rate of −1.2 per cent per annum recently reported for a much larger sample of developing countries (84 in total) during the same period (Lynn and McCarthy, 1989) – a rate influenced by inclusion of countries also included in our sample.

Given the importance of the exchange rate as a policy variable assisting both stabilisation and structural adjustment, the Fund's influence on this instrument is an important finding (see also next section).

The impact on domestic credit, which is a central feature of IMF programme design, will be regarded by it as less satisfactory. There is some reduction in the rate of credit expansion and in the value of credit relative to GDP, but the effects are small and non-significant. Contrary both to expectations and Fund intentions, there is no significant reduction in the share of total credit going to the central government *vis-à-vis* the private sector. More consistent with programme objectives is the significant reduction in budget deficit achieved in Years 1+2 and 3 (although the number of observations in this case is small).

## Persistence

One of the ways in which the tests reported here differ from most of the existing literature is in permitting an examination of programme effects over a longer period, although the small size of our sample should be borne in mind here. It is quite common for such tests in the past to have been confined to the programme year (Year 0 in our terminology) and perhaps the year after. Comparing the Year 0 with the following two columns of Table 3.5, it is evident that tests which are confined to that year would understate the extent to which IMF programmes were associated with changes in the pre-programme situation, for our results for Years

1+2 and even Year 3 are generally larger and more significant than for Year 0.

In fact, the results for Year 0 are generally weak. If we are willing to adopt the language of cause and effect, it appears that programmes in the 1980s were only able to bring quick changes to the exchange rate (devaluations were often 'prior actions') and investment (line 13). A substantially wider range of significant results is obtained for the later years, which is perhaps surprising given that the average programme life in our sample, after allowing for cancellations, was only 18 months. The extent to which BoP improvements persisted into Year 3 is particularly striking.

## Effects of non-implementation

There is a strong presumption that the extent of slippage in policy implementation is greater in the case of uncompleted programmes, for reasons given in the previous section, although some of the policies will have been implemented in virtually all cases, since some measures will have been made preconditions. How much difference does non-completion make?

First, and curiously, it seems that, relative to GDP, credit is restrained *more* in the non-completion cases, overall and with respect to both major components of credit (although the results are statistically non-significant). This raises the questions why the programmes were suspended and whether it was because they incorporated particularly severe, and unattainable, credit ceilings. More in line with expectations, programme non-completion is associated with substantially smaller real exchange-rate depreciations, which suggests that government reluctance to devalue may be a reason for breakdowns, although much of the statistical difference between these sub-groups was the result of Ghana's huge devaluation. Unfortunately, there were insufficient observations on budgetary effects for the results to be disaggregated between completed and non-completed programmes.

Turning from policy instruments to target variables, while the BoP current account result is smaller in the case of uncompleted programmes, such a result is predictable from the worse terms-of-trade experiences of the non-completing countries. As might therefore be expected, the import squeeze is concentrated on the non-completers; their net increase in capital outflows is smaller;

and the overall BoP result is actually larger and more significant than with completed programmes.

As regards the domestic economy, the inflation record is a good deal worse among non-completing countries, with a substantial (though non-significant) rise, against a handsome reduction in the other group. The completers also have a (non-significantly) larger increase in GDP growth. Finally, and puzzlingly, the reduction in total absorption is particularly large and significant in the non-completion cases, with the brunt again mainly falling on investment.

All in all, then, if we focus on the BoP record as the main target variable, it is not obvious that governments which complete their programmes get superior results. We tested for the statistical significance of all the pairs of observations in the 'completed' and 'uncompleted' columns but only the line 5 results on the terms of trade were significantly different, even at a 90 per cent confidence limit. That non-implementation should not have more clearly adverse consequences underscores the limited association between IMF programmes and improvements in economic indicators.

### Effects of past relationships

Finally, although not reported in Table 3.5, we examined whether programme outcomes were correlated with past recourse to the IMF. For this purpose we classified cases into those which were not preceded by a programme, those that were preceded by an uncompleted programme, and those preceded by a fully drawn-down programme. The hypothesis was that countries which had previously worked successfully with the Fund were likely to have less severe programmes and/or to get the more favourable results. There were significant differences among these three categories for a number of our variables, of which the most noteworthy were:

(a) The countries which had previously completed a programme achieved the largest reductions in their current account and overall BoP deficits. They (and to a lesser extent the countries which had previously had a programme but had not completed it) also achieved substantially greater growth in export volumes, which points up the time lags between undertaking adjustment measures and stimulating export performance.

(b) The countries which had not previously had an IMF programme incurred the largest reductions in total absorption (especially in Years 0 to 2) and investment. They also suffered the severest short-term (Year 0) squeeze on credit and import volumes, although this was not sustained in later years.

## EXPERIENCES WITH ESAF[29]

As described in Chapter 2, the Enhanced Structural Adjustment Facility was introduced in 1987 as a temporary facility providing highly concessional assistance in support of medium-term programmes in eligible low-income countries. Being less short-term than stand-bys and more directed to strengthening the supply side of the economy and its institutional structure, ESAF was regarded as more appropriate to the structurally based BoP problems of many low-income countries. It could also be seen as a Fund response to the movement by the World Bank into the arena of macroeconomic policy, with the growth during the 1980s of its structural adjustment lending; and to the increased gravity of the debt problems in the early 1980s, with an associated emergence for the first time of a problem of countries falling into arrears in servicing their past IMF credits.

ESAF was thus intended as a vehicle by which the Fund could adapt its conditionality to the particular problems of low-income countries, keep the initiative *vis-à-vis* the Bank in the area of structural adjustment, and respond to the problems created for heavily indebted countries by the return-flow of its past credits, in effect by refinancing them, and, by doing so, to staunch the growth of arrears.

Great interest thus attaches to the success of ESAF programmes, particularly because of the 'especially vigorous' nature of the policy stipulations attached to them. On their success or otherwise hangs the question of whether the Fund as an institution can operate effectively in the particularly difficult conditions of the low-income countries of Africa and elsewhere. Because ESAF is a rather recent creation, there is as yet little independent evidence on the results of these programmes. However, in 1993 the Fund published a substantial evaluation (Schadler *et al.*, 1993, hereafter referred to as *OP 106*) of experiences with ESAF, and the main purpose here is to appraise this report.

## Nature and conclusions of this study

The study, prepared by members of the Fund's Policy Development and Review Department, examines the experience of the 19 countries that had undertaken ESAF programmes as of mid-1992. These countries, all but four of which were African, between them had had a total of 51 ESAF programmes by that date. The approach is a before–after one, comparing the values of variables in the period before the first adoption of a SAF or ESAF programme against their values after (or in some cases during) the most recent programme. It is particularly concerned with two questions: to what extent were programmes associated with improvements in financial and structural policies, and what were the consequences for external and domestic performance.

Its conclusions are broadly positive (*OP 106*: 39–41). It finds important improvements in economic policies – particularly in the areas of trade and exchange liberalisation, decontrol of agricultural prices and marketing boards, and liberalisation of interest rates – although it adds that in most of the countries reviewed the reforms remain incomplete. It shows faster depreciation of the real exchange rate by comparison with pre-programme years; large reductions in the rate of expansion of domestic credit (although the average growth of money supply was not slowed); and substantial increases in real interest rates. However, in line with evidence presented earlier of limited fiscal impact, progress in strengthening government budgets was patchy: the overall fiscal balance (excluding grants) improved in 12 but worsened in 7 countries.

What about progress towards programme objectives? *OP 106* finds that, on average, ESAF countries have seen improvements in a range of performance indicators, although it adds that there have been disappointments. There was a particularly strong increase in the growth of export volumes, although the current account deficit actually rose from an average of 12.3 per cent of GDP in the pre-programme period to 16.8 per cent in the most recent year (*OP 106*, Table 9), and the report cautions that external viability remains elusive, with visible improvements in only 11 of the 19 countries (p. 29). Interestingly in the light of our earlier discussions of catalytic effects (and rather contrary to Rowlands's findings), net resource inflows (relative to GDP) increased in 8 of the 19 countries but decreased in 7 of them.

These conclusions were welcomed and endorsed by the Fund's Executive Board and written into its 1993 *Annual Report* (pp. 61–4). They may well have played a crucial role in securing the renewal of ESAF in 1994, for agreement on this by the major donor countries is known to have required, in the words of the Managing Director, an 'extraordinary effort' (*IMF Survey*, 21 March 1994, p. 81). However, if indeed donors were swayed by the findings of this report, they may well have been misguided to be so influenced, for reasons to which we now turn.

## A critique

First, we should note that (in common with the previous section) *OP 106* is based on before–after comparisons, but without apparent recognition of the limitations of this method. It also omits any significance testing. Conclusions are drawn on the basis of comparisons of before and after values without consideration of whether the differences lie within the error margins of the values in question. For example, the report states (*OP 106*: 31) that 'most indicators of economic performance improved during the SAF–ESAF period. . . . Real GDP growth, which averaged about 2 per cent a year prior to the respective SAF–ESAF periods, rose to 4 per cent during SAF arrangements and 3 per cent during the ESAF arrangements.' The actual statistics are 2.1 per cent, 4.0 per cent and 2.8 per cent respectively (*OP 106*, Table 9). With 19 observations, the difference between 2.1 and 4 may be significant, but between 2.1 and 2.8? Unfortunately, insufficient data are provided to permit us to test this. The reader is left to infer that differences in outcomes are attributable to programmes rather than to non-programme factors but, in truth, we can only guess at which results are unlikely to have occurred by chance.

The next difficulty concerns differentiation of results according to the vigour of programme implementation. Since implementation is likely to have varied from country to country, it is important to relate outcomes to this factor, as is done in the 1994 World Bank study of its experiences with structural adjustment in Africa (World Bank, 1994b). The Fund study shies away from this, however. Instead, it divides the 19 countries into two groups, according to the progress they made towards BoP viability, and compares these sub-groups (of ten and nine countries respectively) in respect of changes in domestic policies and macroeconomic performance.

One problem immediately suggests itself. As the report spells out with admirable clarity, the concept of external viability is complex and cannot readily be reduced to a single statistic. The authors therefore utilise two different indicators. Even these, they admit, tell only part of the viability story, but when it comes to the comparisons mentioned in the preceding paragraph they confine themselves to trends in debt-service-to-export ratios. However, changes in these may be strongly influenced by non-programme factors, especially debt write-offs and swaps, autonomous changes in export volumes (e.g. as a result of variable weather) and changes in world prices for exports. The report acknowledges most of these possibilities but does not explain what significance remains from a comparison of countries based on changes in debt ratios.

This ducking of the implementation issue is particularly unfortunate because a high proportion of ESAF programmes have taken longer (often much longer) than originally intended. An analysis on data as at end-April 1993 reveals that of the 25 ESAF programmes that should have been completed by that date, only 5 had actually been completed within the originally intended period, a further 14 had been extended and 2 had apparently been abandoned outright. Data on a further 4 were impossible to interpret. Incomplete implementation is a problem for the Fund (a 1991 internal staff review had found that 44 per cent of all ESAF benchmark provisions had not been implemented as agreed) and the authors, of course, had access to country-by-country information about this that would have permitted them to control for this factor. Neglect of this is doubly unfortunate, for we have seen in the previous section that for the programmes studied there was no strong connection between implementation and BoP results.

Leaving that to one side, the rationale for comparisons of the two groups in respect of the extent of their policy reforms is clear: if the policies are effective, greater action on policy should lead to more progress towards viability. Unfortunately, the evidence for this turns out to be weak, as the report acknowledges (*OP 106*: 37):

> Generally, the countries with the best external and domestic performance undertook stronger financial adjustment measures than those where progress towards external viability and domestic performance was weaker. However, except in the area of fiscal policy, differences were not great.

Even in the case of fiscal policy the differences seem less than 'great'. The following figures record for the two sub-groups central government budget deficits as a percentage of GDP (from *OP 106*, Table 12):

| | *Three years before programme* | *Most recent year* |
|---|---|---|
| *Countries making more progress towards viability* | | |
| Median | −9.1 | −3.0 |
| Mean | −12.6 | −5.7 |
| *Countries making less progress towards viability* | | |
| Median | −7.2 | −4.1 |
| Mean | −8.2 | −3.9 |

The desirability of statistical testing is self-evident. In this case, enough information is presented to permit it, so we undertook a standard two-tailed test. Unfortunately this indicated that the 'great' differences in the means of the two groups fell well short of normally acceptable significance levels.[30] We might recall here that Nashashibi *et al.* (1992) found only a non-significant tendency for SAF/ESAF programmes to be associated with improved fiscal balances. Remarkably, *OP 106* makes no reference to the Nashashibi study.

However, there is another significant-looking policy difference between the two groups which is not at all what the authors would have wanted to find. These are figures for average annual percentage depreciation of the real exchange rate during the period of the ESAF programme (*OP 106*):

| *Countries making more progress towards viability* | | *Countries making less progress towards viability* | |
|---|---|---|---|
| Median | −1.8 | Median | −3.9 |
| Mean | −2.4 | Mean | −8.6 |

The larger devaluations are associated with less BoP improvement. The authors assert that this is explained by the worse terms-of-trade experiences of the non-progressing group during the programme period, but they do not demonstrate this. Nor do they contemplate that the fiscal differences may also have been affected by this factor, not least because most fiscal improvement is shown to have come from the revenue side.

In short, the evidence presented provides little support for the claim that those which secured the greatest progress towards viability undertook the strongest measures. What now of the domestic economic performance of the two sub-groups? On this the authors are clear (*OP 106*: 34): 'Those countries that made the most progress in improving their debt position were also those that showed the strongest improvement in their domestic economic performance.'

Here the evidence seems stronger, as indicated by GDP growth rates (*OP 106*, Table 11):

| | *Three years before programme* | *Most recent year* |
|---|---|---|
| *Countries making more progress towards viability* | | |
| Median | 1.4 | 4.0 |
| Mean | 1.3 | 3.7 |
| *Countries making less progress towards viability* | | |
| Median | 3.0 | 2.5 |
| Mean | 3.3 | 1.9 |

(Comparisons of reductions in inflation are also substantially to the advantage of the 'progressors'.) But what light do such results throw on programme effectiveness? Not much, it appears, unless we can regard the 'progressors' as 'strong implementers'. As we have seen, the evidence for doing so is slight but the authors seem to view things that way.[31]

The truth is that external influences emerge as the dominant influence on the comparative performances of the two sub-groups.[32] The report shows that the terms-of-trade experiences of the 'non-progressors' were a good deal worse than their comparators, and that the former received much smaller increases in net financial resource transfers. Not surprisingly, therefore, they suffered far more import compression (annual percentage changes in import volumes, from *OP 106*, Table 11):

| | *Three years before programme* | *Most recent year* |
|---|---|---|
| *Countries making more progress towards viability* | | |
| Median | 1.8 | 2.2 |
| Mean | 1.5 | 3.6 |
| *Countries making less progress towards viability* | | |
| Median | −2.7 | −9.2 |
| Mean | −2.7 | −8.9 |

Within the methodology of this report it is impossible to disentangle the influences of policy, resource flows and other factors affecting economic performance. But it is easy to read the above comparisons of the domestic economic performances of the two groups of countries as simply another confirmation of the binding nature of the foreign-exchange constraint. It is hazardous to infer *anything* about programme success from evidence of this kind.

There is a further problem of interpretation, of some embarrassment to the Fund. The report rightly stresses the highly adverse conditions which afflicted the 19 countries prior to their SAF–ESAF period (*OP 106*: 4):

> By the mid-1980s the interplay among the weak productive bases, periodic disruptions from poor weather and civil strife, sharp deteriorations in the terms of trade, and inadequate policy responses to emerging problems had left most of the countries with low growth and large external imbalances. . . . In all but five countries, real GDP per capita had fallen during the three or more years preceding the first SAF or ESAF arrangement. Often, the most pronounced weakness was in the export sector. Absorption, however, was kept high by large public sector deficits financed by bank credits and external inflows.

The point the authors rightly want to make is that it would be unreasonable to expect rapid progress on the basis of such a dire starting point. However, we learn later that, 'Most of the 19 countries had begun the process of reform and macroeconomic stabilization – often with IMF support – prior to their SAF–ESAF arrangements' (*OP 106*: 7), i.e. through stand-by programmes. Only two of the 19 were embarking upon IMF programmes for the first time, although some of the others were lapsed reformers.

This is interesting information. 'Before–after' tests usually compare situations before and after adoption of adjustment measures. In the present case, however, to a substantial extent the comparison is between the results of stand-bys and of SAF–ESAF programmes. Given the gravity of the pre-SAF–ESAF situation just described, the inference suggests itself that stand-by programmes had been markedly unsuccessful. It is only a partial exaggeration to say that the before–after comparisons in *OP 106* are testing whether SAF–ESAFs work better than stand-bys. Unfortunately, the report is too flawed to permit a conclusion on this.

## Other observations

### *Comparisons with World Bank results*

The structural adjustment movement has brought the IMF and World Bank closer together. By coincidence, they both produced evaluations of their experiences in low-income countries at about the same time (confined in the case of the World Bank (1994b) report to sub-Saharan Africa). This is not the place for a systematic comparison of the two. Neither is able to make any large claims for the results of its programmes. What is striking, however, is the absence of consensus between them on specific countries. The Bank report (1994b: 58) identifies six countries which it regards as showing large improvements in macroeconomic policies. Of these, only three are included in the Fund study, of which Gambia and Ghana are classified by the Fund as having made relatively good progress towards viability but the other (Tanzania) is placed in the non-improving group. More contrasts emerge among the countries judged by the Bank as having had deteriorating macroeconomic policies. Only two of these are also covered by the Fund report – and both of these (Mozambique and Togo) are classified as 'progressors'.

Once again, we are reminded of the fragility of the results of programme evaluations. And then there is . . .

### *The sad case of the evaporating growth objective*

We showed in Chapter 2 that publicity about ESAF gives prominence to the objective of fostering 'adjustment with growth', as contrasted with the more exclusive concentration on BoP viability in stand-by programmes. Unfortunately, the present report gives the lie to this apparent re-ordering of priorities. True, on page 9 it suggests that the need for stabilisation in ESAF countries was not as dire as in other countries, but by the next page this is forgotten, so that we read of 'the *imperative* of stabilizing the economy' (italics added) in the 19 countries under study (*OP 106*: 10).

More to the point, the overwhelming emphasis in *OP 106*'s evaluation is on progress towards external viability; the growth record gets scant attention and the relevance of the growth criterion is implicitly rejected. This impression is reinforced by the fuller treatment in the pre-publication version of the report of the extent to which medium-term programme objectives had been achieved (IMF, 1993: Annex II). These are stated in terms of levels of imports, exports and reserves, and the size of the current account deficit. Growth is simply not considered, nor any other indicator of domestic economic performance.

There is another feature of this. Earlier we quoted the report's finding (*OP 106*: 31) that 'Real GDP growth, which averaged about 2 per cent a year prior to the respective SAF–ESAF periods, rose to 4 percent during SAF arrangements and 3 per cent during the ESAF arrangements.' Note the formulation, 'rose to . . . 3 per cent'. In fact, mean growth rates are shown as *falling* from 4.0 to 2.8 per cent between (low-conditionality) SAF programme periods and 'especially vigorous' ESAF programme periods (in all except one of the 19 countries an ESAF programme was preceded by a SAF programme). If these figures mean what they appear to mean, ESAF programmes which follow SAF programmes *depress* growth. We might also note that a 2.8 per cent growth rate is the same as the average population growth of these countries, implying static per capita incomes.[33]

These inconvenient results are ignored in the text. The standard Fund response would be to appeal to the full wording of the ESAF guidelines, which talk of '*creating the conditions* to achieve sustained economic growth' (italics added) and to urge that ESAF programmes are indeed creating those conditions. The difficulty with this defence, however, is that it is also the one the Fund has always used with respect to its conventional stand-by programmes, which only goes to underline the lack of difference between ESAF and stand-by programmes in this regard. The report's authors have also responded that, because of time lags and other reasons, it would not have been appropriate in this report to look for evidence on growth effects, but this is a defence which sits ill with the prominence it gives to the conclusion, just quoted, that ESAF countries experienced improved growth.

## Conclusion

What, then, might we conclude about results from the Fund's ESAF experiment and its ability to operate effectively in low-income countries? Sadly, it is hard to conclude anything on the basis of *OP 106*. While making few explicit claims, the authors leave the impression that results have been reasonably satisfactory, and the Fund's 1993 *Annual Report* concluded from this that ESAF 'had proved an effective mechanism for Fund involvement in low-income countries' (p. 61). Unfortunately, the report cannot sustain such a conclusion. We can accept, and there is supporting evidence elsewhere, that there have been policy improvements in the countries studied, in directions of which the Fund would approve. We can (just) accept that there is evidence, properly qualified in the report, of some improvement in external performance (particularly with regard to growth in export volume), and perhaps in domestic economic performance too.

What we cannot do is believe that *OP 106* has demonstrated any systematic connection between these changes and the implementation of ESAF policies. There is too much methodological confusion for any such conclusion to be drawn. What we can more safely infer is that the foreign-exchange constraint remains binding in many low-income countries (so the Fund's focus on the BoP does at least address the area of most crucial importance) and that movements in the terms of trade continue to have a potent influence on their abilities to strengthen their economies. We are also entitled to adduce that the ESAF's stated growth objective is more presentational than operational, although by addressing deficit countries' structural weaknesses the ESAF *should* have a positive longer-term growth effect. Finally, we can note that low-conditionality SAF programmes appear to be associated with better (but statistically significant?) growth records but less BoP progress than high-conditionality ESAFs.

Overall, the inconclusive nature of this study is much to be regretted. Both the Fund and borrowing countries *need* its structural adjustment lending to be a success, as well as firmly based evidence on which future reforms can be designed.

# A SURVEY OF COUNTRY EXPERIENCES WITH IMF PROGRAMMES

## Introduction

So far we have concentrated on the results of quantitative cross-country assessments. We turn now to present the more qualitative findings of country-specific case studies. It was suggested in the first section of this chapter that a well-conducted case study has considerable advantages over multi-country approaches in virtually all respects save one: the ability to arrive at generalisable conclusions. In order to overcome this limitation, the results which follow are based on a survey of studies relating to the experiences of 17 developing countries which adopted IMF programmes in the 1980s.

After a brief explanation of the approach used and of the nature of the country sample, we then discuss the light which the country studies throw upon the circumstances leading to adoption of a programme. Next, evidence relating to the cost effectiveness of the programmes adopted is presented, and finally the discussion above of the determinants of programme effectiveness is taken further.

## Approach

The first step in this research was to design a standard format-cum-checklist within which country materials should be organised. Next, this framework was tried out on a small number of countries, chosen because they were fairly familiar and known to have reasonably extensive literatures. This showed the framework to provide satisfactory guidance, and the next phase was to conduct a literature survey to establish a list of developing countries for which there was a published or 'grey' literature studying their experiences with IMF programmes.

A number of these countries were then selected at random and the order in which they were drawn was noted. We then worked down this list up to the limit of the research time available. The resulting coverage could not be described as a random sample since it was biased towards those countries for which there was an (English-language) literature. However, it can be seen that procedures were incorporated to safeguard against researcher bias.

Including those covered by the pilot studies, there was a total of

36 countries on the list, of which it was possible to complete surveys of 17, which between them started 48 programmes in the period 1979–89. However, the available literature only covered a proportion of the total of 48 programmes. The countries were: Bangladesh, Brazil, Costa Rica, Côte d'Ivoire, Dominica, Dominican Republic, Gambia, Ghana, Jamaica, Malawi, Mexico, Morocco, Pakistan, Philippines, Somalia, Sudan, Tanzania.

In terms of regional coverage, type of economy and other characteristics, this group seemed reasonably representative, although Latin America was somewhat over-represented. However, two substantial biases were discovered. Comparing the experiences of the 17 countries with all those which adopted IMF programmes in the 1980s, our sample included too high a proportion of EFF programmes – 33 per cent as against only 11 per cent for the total number of programme countries. Partly as a result of this, our sample also contained a disproportionately high share of programmes which broke down before the end of their intended life – 63 per cent against 50 per cent. This bias may perhaps result from the special interest of researchers in the EFF experiment; and from a tendency for programmes which failed, in the sense of having been abandoned, to provoke more of a research response than programmes that went smoothly. However, we believe that the uses to which the country information is put in what follows minimise the possible distorting effects of these biases.

## Sources of destabilisation

Much of the controversy about the role of the IMF in developing countries has centred around the question of whether their BoP and other macroeconomic problems are typically the result of domestic mismanagement or 'exogenous shocks' emanating from the world economy or other factors beyond the control of national policy-makers.[34] Critics of present arrangements argue that Fund-type programmes are most appropriate for countries whose macro problems are largely the result of policy mismanagement leading to excess aggregate domestic demand, but that in the turbulent conditions of the world economy since the early 1970s developing countries' problems have more typically been the result of external shocks. Where that is the case, it is argued, an enlarged flow of international assistance is the more appropriate policy response, i.e. finance rather than adjustment.

Fund staff more frequently blame macro problems on domestic mismanagement and argue that, in any case, countries have to adjust to shocks, since the industrial countries are unwilling to provide the additional financing that would be necessary if adjustment were to be avoided. To this it can be responded that, even though adjustment may be unavoidable, it is important to diagnose the sources of BoP difficulty in order to design the most appropriate policy response (this latter argument is developed further in *The Quest*, Chapter 8).

What light do our country studies throw on this set of issues?

### *The country evidence*

Judgements on how the balance should be struck are unavoidably subjective, but in a majority of our countries (10 out of 16 for which the literature permitted a judgement) the circumstances leading to the adoption of Fund programmes appeared to be an admixture of exogenous shocks and poor domestic macroeconomic management, although the nature of the shocks and policy weaknesses naturally varied.

The Gambia provides a well-documented illustration. In the decade to the mid-1970s impressive economic progress had been achieved, while maintaining approximate BoP equilibrium. Then a number of adverse external developments intervened: the first two oil shocks; deteriorating terms of trade (which fell by 25 per cent in 1976/77–1984/85); the Sahelian drought of 1981–5. These, however, interacted with a weak policy framework. Government expenditures were allowed to grow faster than revenues so that by 1981 the budget deficit equalled 11 per cent of GDP. This and growing BoP difficulties led to extensive borrowing from abroad and a rising debt burden. Currency overvaluation led to the emergence of a parallel foreign-exchange market. Such factors led to a further 'external shock' in the form of declining aid receipts. Private investment collapsed; export volumes fell steeply (much worsened by smuggling of groundnuts into Senegal); foreign exchange reserves fell to the equivalent of two weeks' imports. There was a further exogenous factor peculiar to the country's position: a coup attempt in 1981 which the government could put down only with military help from Senegal, and this then led to uncertainty about the Gambia's future as a sovereign state. This helped to convince the Gambian government that radical policy changes were needed.

Other cases illustrate a similar mix of domestic and external causes. Costa Rica suffered from the oil shocks, a related deterioration in its terms of trade (which fell by a third in 1977–81), and – having borrowed substantially from commercial banks on variable-rate terms – from the rise in world interest rates in the early 1980s. However, the harmful effects of these developments were aggravated by expansionist fiscal policies and consequentially large pressures of excess demand, leading to an inflation rate which approached triple-digit levels in 1982, and by a seriously overvalued currency (necessitating a devaluation from 10.9 colones per SDR in 1980 to 42 in 1981).

Jamaica is another example, although there is much controversy about where the balance should be struck between exogenous shocks and domestic weaknesses. Here the relative lack of interest on the part of the People's National Party government in the precepts of macro management in a highly vulnerable economy, and its excessive faith in import substitution, magnified the effects of the oil shocks and the weakness of the world aluminium market, and led not only to a major economic crisis but also to an 'IMF election' which unseated the PNP.

The Philippines provides an Asian example. The well-known inefficiencies and excesses of the Marcos administration, and the resulting economic mismanagement, have tended to mask the seriousness of the external shocks which the economy of the Philippines experienced in the early 1980s, where the effects of rising oil prices and interest rates were compounded by declining demand for some of the country's manufactured exports (see UNICEF Manila, 1988: Table 7.2).

We can also mention Somalia, a country with an extremely fragile economy where domestic policy-makers, already struggling with large budget deficits, losses of international reserves and rapid inflation, were quite unable to cope with a situation in which the oil and interest-rate shocks were aggravated by drought and then, in 1983, by a Saudi Arabian decision to ban imports of cattle from Somalia, which previously had contributed 75 per cent of the country's total export earnings.[35]

The second largest group in our sample – four in number – are countries where it seems most appropriate to attribute their BoP and other macro problems essentially to domestic policy weaknesses. Of course, these economies were not immune to adverse external shocks, but the essence of their problems seems domestic in the sense that the shocks merely hastened an economic crisis that appeared inevitable anyway.

This group includes the two largest economies in our sample: Brazil and Mexico. In the former case, access to international bank credit was used in the 1970s and into the 1980s as an alternative to policies that would adjust the economy to the oil shocks. Expansionary fiscal and monetary policies were pursued in the face of already substantial macro imbalances. Large-scale external borrowing could not prevent a rapid haemorrhage of foreign-exchange reserves. Although the currency was allowed to depreciate, this was not sufficient to compensate for the rapid inflation (averaging around 100 per cent per annum in the period 1979–82), and the real exchange rate appreciated. The rise in world interest rates and weakening world demand for the country's exports merely executed the *coup de grâce*, precipitating in 1982–3 a crisis of macroeconomic management which in any case had seemed only a matter of time.

A not dissimilar story could be told about Mexico, with the major exception that this country was a net beneficiary of the oil shocks. The diagnosis here is that the government failed to understand the Dutch-Disease problems created for the rest of the tradable-goods sector by the rapid growth of the oil industry, and misjudged what proved to be temporary rises in real oil prices as a permanent shift (although it was in plentiful company in making this mistake). Here too there were major current account BoP deficits, a large-scale recourse to short-term bank borrowing and an appreciating real exchange rate. The problems were aggravated by a flight of capital from a private sector which did not share its government's confidence. Once again, rising world interest rates, coupled with falling real oil prices, tipped an unbalanced economy into crisis.

Africa provides our other two examples. In Ghana two decades of generally low-quality economic policies had brought the economy to the point where living standards were falling drastically, there was a retreat from the modern economy back to subsistence production, the parallel market rate for foreign exchange was 30–40 times the official exchange rate, the tax base had been seriously eroded, and the government's budget was largely out of control, there was rapid inflation and acute shortages of imported goods. Here the *coup de grâce* was administered by a serious drought and a huge influx of nationals deported from Nigeria, forcing the government to the realisation that it had no alternative but to turn to the IMF and World Bank.

In Sudan too the weakness of past policies had left the economy extremely vulnerable to the slightest ill wind blowing from abroad. Ambitious and undiscriminating investments in infrastructure had led to many low-productivity public sector investments, and this combined with a diminishing revenue base to create a rapidly widening budget deficit. There was substantial inflation, an overvalued currency and large BoP strains leading to import shortages. Above all, there was a government which displayed little interest in macroeconomic management.

Finally, we come to two countries where the origins of the BoP difficulties appear essentially exogenous, in the sense that there is no reason to believe that policy weaknesses would otherwise have created a macro crisis: Bangladesh and Dominica. In the former case, Matin (1986) argues that prior to 1979 the economy was in apparently good shape, with fairly rapid economic growth, low inflation, rising aid receipts and a reasonably comfortable BoP situation (although there were fiscal strains). What then forced the government to ask for IMF support in 1979 was a serious drought and abnormal food import requirements. In Dominica, the key events leading to the 1980 EFF programme were devastating hurricanes in 1979 and 1980, and a phasing-out of UK budgetary support following independence in 1978, superimposed on a tiny, extremely fragile economy.

### *Points arising*

The country evidence draws attention to the importance of considerations which hitherto have been rather neglected and which the case sketches above do not fully bring out.

First, we are struck by the frequency with which *natural disasters* are an important, sometimes dominant, factor in decisions to adopt an IMF programme. The drought in Bangladesh has just been mentioned; droughts also had a major influence in the Gambia, Tanzania and Ghana, where they led to forest fires and extensive agricultural damage. Severe hurricanes were important causes of the crises in both Dominica and the Dominican Republic.

Second, we are impressed by the extent to which *dependence on one or a few primary product exports* added to the management problems of our sample of countries. We are particularly struck by the frequency with which *commodity booms* are mismanaged, so that what ought to be a windfall gain leads the economy directly into a macroeconomic crisis. The history of Côte d'Ivoire provides a case in point, leading to its adoption of an EFF programme in 1981. Here the 1976–7 boom in world beverage prices (largely the result of severe frosts in Brazil and therefore obviously temporary) led to the accrual of large windfall gains to a government 'stabilisation fund' which, instead of being set aside for later periods of price weakness, were used to finance a major public investment programme. State expenditure levels rose and were not scaled back again when the boom duly ended in 1978, leaving a budget deficit equivalent to about 10 per cent of GDP in 1979–80, substantial inflationary pressures and large losses of reserves.

This is by no means an isolated phenomenon. It has been well documented elsewhere for other countries (see Garcia and Llamas, 1988; World Bank, 1988: 71–4; Bevan *et al.*, 1986). Among our sample there were comparable instances in the Dominican Republic, where a 1979–81 surge in world sugar prices was allowed to mask the underlying weaknesses of the economic situation; in Mexico, where we have already referred to the way in which the 1972–3 and 1979–80 oil price rises contributed to the macro imbalances which came to the fore in 1982–3; and in Morocco, where a short-lived quadrupling of the world price of its major export, phosphate, left a legacy of unsustainable public expenditure levels.

Quite apart from any tendency to mismanage, or misread, temporary commodity booms, it is evident from our cases that those heavily dependent on one or a few primary products are especially vulnerable to BoP crises, to which recourse to the IMF becomes the only perceived solution. Apart from those just mentioned, we could add Dominica, the Gambia, Ghana, Jamaica, Malawi, Somalia, Sudan and Tanzania (i.e. a substantial majority of the total sample) as countries adversely affected by such dependence during the 1980s.

Such experiences have important policy implications. One, obviously, is to warn governments against a myopic and ultimately destabilising misuse of temporary export booms as devices for raising investment or consumption to levels which cannot thereafter be sustained – and for the IMF to be vigorous in offering advice along these lines. The adverse effects of commodity dependence, *per se*, point additionally to the desirability of export diversification policies. This is not merely a policy indication for national governments; it affects the IMF too. Where commodity dependence is an important source of BoP weakness, policies of export diversification should also be incorporated in IMF programme design. While the Fund can point out that it does frequently incorporate currency devaluations in its programme designs and that such action is necessary for export diversification, it much less frequently goes beyond the exchange rate to other pro-diversification measures.

Finally under this heading, we can note the frequency with which *slowness in policy response* was an important element leading up to the necessity for an IMF programme. Thus, for Tanzania, Ndulu (1987) draws attention to a long lag before domestic policies began to grapple with the problems created in the late 1970s and early 1980s by deteriorating terms of trade, the heavy costs of the war in Uganda and adverse weather from 1979–81, exacerbated by disastrously bad storage of agricultural surpluses grown in the period 1976–8. We referred earlier to a similar tardiness of policy response in the Gambia, where matters had to come to crisis point before there was a serious policy response. Precisely the same complaint could be made about Ghana up to 1982. A related point could be made about Brazil, Costa Rica and Mexico: that they preferred running up large external debts to more timely policy actions to restore balance to their economies. Morocco was another country which bought time by borrowing and which was slow to react to economic deterioration and the impact of external shocks. Much the same goes for the Philippines and Sudan.

## Evidence on programme cost-effectiveness

Our country survey throws additional light on various factors bearing upon the cost-effectiveness of Fund programmes, including consequences for income distribution and political stability, and possible tensions between short-run IMF-style stabilisation and longer-term programmes of structural adjustment. We begin, however, by examining what the effects are of frequent, as contrasted with occasional, recourse to Fund credits, for our sample includes several countries which had a succession of programmes during the 1980s.

### *The effects of a long-run Fund presence*

The countries analysed here are as follows, with the figures in parentheses indicating respectively the number of programmes and the total number of years in which these were in force (i.e. not suspended or abandoned) during 1979–89, which is our period of analysis:[36] Bangladesh (5, 7½), Costa Rica (6, 7½), Côte d'Ivoire (5, 8), The Gambia (6, 6½), Ghana (6, 5½), Jamaica (6, 9½), Malawi (4, 6½), Morocco (7, 8).

The question we ask here is, what was the state of these countries' economies, particularly their BoP, at the end of a decade in which they made such extensive use of Fund support? We approach this by briefly surveying the records of these countries, beginning with the worst case and moving in roughly ascending order of achievement.[37]

The only unambiguous failure was Côte d'Ivoire. Despite having programmes in operation during eight of the eleven years under examination, it ended the decade in economic crisis. There was a huge deterioration in the current and overall BoP in the period 1985–9; there was a marked adverse trend on long-term capital account; export volumes were static; the real exchange rate was appreciating, being by 1989 nearly 30 per cent above the 1985 level; the country needed recourse to repeated debt reschedulings. There was also a serious worsening in other aspects of economic performance: the domestic savings ratio declined steeply and the investment ratio was also on a falling trend, ending the decade at only 10 per cent; the government's budgetary situation deteriorated in the later years; *per capita* incomes were declining rapidly by the end of the period. Indeed, by then the economy was in deep crisis.

One other feature distinguishes Côte d'Ivoire: a fixed nominal exchange rate, determined by membership of the Franc Zone monetary union, which between 1948 and 1993 maintained an unaltered exchange rate between the CFA and French francs. While Côte d'Ivoire, along with other member countries, had doubtless benefited from the convertibility and financial stability afforded by the Franc Zone, this led to an appreciation in the country's real exchange rate during the period under study. It is tempting to see the government's – and the Fund's – inability to use the exchange rate as a major reason why Côte d'Ivoire ran into such large macroeconomic difficulties.

The problem is probably more deep-seated than this, however. The country was, in fact, able to sustain a trade surplus throughout the period, with the chief BoP difficulties arising from the claims of debt servicing and an important outflow under the heading of 'private unrequited transfers'. As we have already suggested, fiscal discipline left much to be desired. More pervasively, it is widely held that, as President Houphuët Boigny's command diminished during the 1980s, economic policies became more erratic and less compatible with macroeconomic stability.

The position of Malawi was also rather unsatisfactory at the end of a long period of support by the IMF, despite superficial signs of improvement. After a considerable period of stagnation, export volumes were tending to increase; the ratio of the current account deficit to GDP started the period at a full 25 per cent of GDP and was down to around 4 per cent by 1988. The net long-term capital account showed some improvement and the overall balance was in surplus in 1987–8, with sizeable replenishments of reserves. To a substantial extent, however, the BoP improvements were achieved by cutting imports. There was also a worrisome tendency for the real exchange rate to appreciate towards the end of the period.

Malawi's domestic performance indicators are hard to read, because they were highly unstable. The inflation rate rose consistently from 10.5 per cent in 1985 to nearly 34 per cent in 1988, but then fell abruptly to 12 per cent. Similarly with economic growth: this fell from 4.5 per cent in 1985 to about 1 per cent a year later and then gradually climbed back to over 4 per cent by 1989. Abstracting from these fluctuations, per capita incomes remained at best static. This was not surprising since the economy was subjected to considerable international difficulties and gross fixed investment had been on a declining trend. Domestic saving was declining steeply. Fiscal trends were erratic, although the public finances ended the decade better than they began. This, then, is a difficult record to evaluate. There were improvements in some BoP and other indicators, despite adverse circumstances, but deteriorations in other aspects. Overall, the position was unsatisfactory and fragile.

Bangladesh also had a mixed record. Export earnings grew well, if erratically; the current account deficit in most of the later years of the period was smaller relative to GDP than at the beginning, but jumped in 1989 (following floods and consequent falls in exports and increased import needs); the overall BoP was generally in surplus. Inflation was roughly stable at about 10 per cent throughout. More seriously, estimated gross saving was down to a paltry 1 per cent of GDP by 1989 and fixed investment was under 12 per cent. Per capita incomes were generally static, with GDP growing at only about 2 per cent per annum in the period 1987–9. Fiscal data are poor but suggest that the public finances had weakened during the decade. Overall, it is difficult to assert that the always weak Bangladeshi economy strengthened during these ten years.

There is also ambiguity in Jamaica's record. BoP indicators fluctuated rather widely during the latter part of the period and it was difficult to discern clear trends. Presumably in response to a large exchange-rate depreciation in the mid-1980s, export volumes had reversed their previously declining trend, and there were also indications of diversification. The current account deficit first diminished and was turned into a surplus in 1988, although there was a large deterioration in 1989, due to disruptions caused by Hurricane Gilbert. With long-term capital flows, international reserves and the overall BoP fluctuating sharply year to year, and with a continuing need for frequent recourse to debt reschedulings, it is difficult to assess the underlying state of the BoP, although it could hardly be described as secure.

At home, the inflation rate fluctuated, lower at the end than at the beginning but tending to accelerate. Gross fixed investment remained fairly steady at around 20 per cent, but the data cease with 1987. Fiscal data are also poor but indicate an improving trend. Although there had been large short-term fluctuations and some recovery from the recession of the earlier part of the decade, the economy remained sluggish, with population growth still exceeding GDP expansion in most of the later years.

The effects of Hurricane Gilbert and the general instability of economic indicators make it more than normally difficult to assess the condition of the Jamaican economy at the end of the decade. There was neither an obvious strengthening nor a clear deterioration. The economy remained highly fragile, and the BoP could easily deteriorate again. Also fragile were relations with the IMF, with successive governments being critical of the Fund from time to time, substantial slippage in the implementation of most programmes, and three of six programmes in the period cancelled because of non-compliance.

A firmer favourable judgement can be made about Costa Rica, with improving indicators on both BoP and domestic indicators. There were deficits on the BoP current account throughout the period but, relative to GDP, by the final years these were only about half of the levels experienced at the beginning. Despite fluctuations on capital account, the reserves were rebuilt. Export volumes increased, aided by a substantial depreciation in the real exchange rate. Debt ratios were tending to decline but frequent reschedulings were still necessary. The domestic economy was, moreover, fairly buoyant, with GDP growth of around 5 per cent in the last years of the period. The investment rate was fairly high, approaching 20 per cent of GDP in the later years. Having been brought down from a high of 90 per cent in 1982, the inflation rate showed some tendency to rise again, being in the 15–20 per cent range for the final three years. Fiscal data indicate only small overall deficits in recent years – a major improvement on the large deficits at the beginning of the decade. All in all, the economy appeared stronger by the end of the period.

This judgement also applies to the Gambia. Although this country's database is weak, the statistics show improvements in the overall BoP, from large deficits in the earlier years to substantial surpluses in the concluding years, permitting reserves to be rebuilt.[38] This improvement was, moreover, achieved in the face of worsening terms of trade, although it was greatly facilitated by enlarged aid receipts. The real exchange rate depreciated until 1986 but then tended to drift up again. The export sector remained completely dominated by groundnuts (making up 90 per cent of earnings, on average) and its performance was therefore strongly influenced by changes in weather conditions.

After a decade of declining *per capita* incomes to the mid-1980s, economic growth averaged about 6 per cent per annum from 1986–9 and inflation was down, although still well above 10 per cent. Fiscal trends were also positive, with substantial improvements on current and overall balances, reflecting improved expenditure control and larger revenue ratios. Investment was stable at around 25 per cent of GDP over the decade, while the domestic saving ratio was much improved – from a very low base. Overall, the Gambian economy has been on an improving trend, although much of this was on the basis of a mono-crop export base and considerably enlarged receipts of external support.

This brings us to Ghana, whose apparent successes with its

adjustment programmes since 1983 have attracted much international attention. Does our evidence bear out this success story? Yes, but with qualifications. Although the BoP current account has remained substantially in deficit, this has partly been due to badly needed increases in imports and has been financed by a major growth in capital receipts. The overall balance was in surplus for most of the later years and there was a much-needed rebuilding of reserves. Moreover, this record was achieved in the face of strongly adverse movements in the terms of trade throughout most of the period. Underlying this record was a very large depreciation of the real exchange rate from an initially grossly overvalued level, and large increases in the volumes of some exports.

At home, too, the Ghanaian economy has been recovering from the nadir to which it had sunk by 1982, with the economy growing annually at around 5 per cent from the mid-1980s to the beginning of the 1990s (aided in the early programme years by a return to more normal rainfall levels). The budgetary situation has improved, with increases in revenues relative to GDP permitting both higher expenditures and a move from large, inflationary budget deficits to small surpluses in the final years of the period.

The qualifications? Inflation remained persistently high, averaging 30 per cent per annum in the period 1986–9, even though well down from the triple-digit levels of pre-programme years. Second, Ghana's recovery, like that of the Gambia, was based on abnormally high levels of external assistance, making it especially difficult to disentangle the effects of the aid from programme results. Finally, capital formation, especially in the private sector, remained depressed, with a total investment ratio estimated in the range of 6–12 per cent in 1988. Ghana's impressive turn-around remained fragile.

Of all the countries surveyed here, Morocco's record was arguably the most impressive. Most of its BoP indicators were strengthened, although aided by favourable terms of trade in the second half of the period and substantial external assistance. After a succession of large deficits, the current account moved into surplus from 1987–8 (although it slipped back rather severely in 1989); there were surpluses on the overall BoP, despite reduced inflows of long-term capital. A gradual real depreciation of the exchange rate presumably increased the competitiveness and/or profitability of exporting, and export volumes rose steadily during the period 1982–8. There was, however, a deterioration in most indicators in 1989. There were frequent debt reschedulings and debt-servicing burdens remained large, although they declined towards the end of the period.

In the domestic economy, the inflation record was also satisfactory, with a reduction from 8–10 per cent per annum during 1982–6 to around 3 per cent during 1987–9. In the face of a flat tax ratio, government expenditure ratios were gradually scaled back, leaving an overall budget deficit under 5 per cent of GDP in 1988 (latest), compared with 10 per cent in the early 1980s. Although there were considerable year-to-year fluctuations, economic growth was generally well in excess of increases in population in the later years. Fixed investment was sustained at over 20 per cent of GDP throughout, although the ratio tended to drift down a little; the savings ratio improved. Overall, this was an economy which clearly turned around during the decade, an achievement all the more impressive because it did not rest upon abnormal inflows of aid monies.

An attempt to summarise the foregoing is set out in Table 3.6. What this suggests is that a majority of these eight frequent adopters of IMF programmes experienced a strengthened BoP situation in varying degrees, although the record on domestic economic performance was more variable, with unambiguous improvements in only a few cases.

Of course, causality cannot be automatically inferred from association between frequency of Fund use and economic performance. In particular, an improving outcome might be regarded as predictable because governments which most frequently enter into, and complete, Fund programmes are liable to be among those most concerned with the general quality of their economic policies and most likely, therefore, to achieve improved economic outcomes. Consistent with this, most of the countries also undertook programmes of 'structural adjustment' with the support of the World Bank, which creates the additional complication that it is difficult and artificial to try to separate out the economic effects of the IMF's stabilisation programmes from those of structural adjustment measures. Nevertheless, the Fund can reasonably derive satisfaction from the story told here, even if the evidence remains inconclusive.

Our discussion has so far mainly concerned the benefits that may be derived from IMF programmes. We turn now to some of the costs that may be incurred in adopting such programmes.

*Table 3.6* End-of-period (1989) position of IMF frequent users

| | *Bangladesh* | *Costa Rica* | *Côte d'Ivoire* | *Gambia* | *Ghana* | *Jamaica* | *Malawi* | *Morocco* |
|---|---|---|---|---|---|---|---|---|
| Balance of payments | Export earnings growing. Also long-term capital inflows. RXR[a] depreciated. Current a/c deficit growing but overall balance generally sound. | Export and import volumes on rising trend. Long-term capital inflows reviving but still small. Current a/c deficit/GDP down. Reserves rebuilt. | Import volumes down, exports flat. Reduced capital inflows. Appreciating RXR. Large, growing current a/c deficits. Overall balance worsening. Debt difficulties. | Improvements in current and overall balances but much of it on basis of increased aid. RXR depreciated. | Overall BoP improved despite worsening terms of trade but dependent on large aid receipts. RXR depreciated. | RXR depreciated and exports reviving. But BoP indicators fluctuate; trend uncertain. Substantial debt, frequent reschedulings. | Export volumes rising. Current a/c deficit/GDP lower than at beginning but rising despite import squeeze. RXR depreciated but rising. | Export performance and BoP balances generally improving. Also terms of trade. RXR depreciating. |
| Inflation | Stable at *c.* 10%. | Reduced but still 16% in 1989. | Moderate throughout. No trend. | Down in final years. | Down but still *c.* 30% in final years. | Fluctuating. | Fluctuating. No clear trend. | Reduced to *c.* 3%. |

| | | | | | | | | |
|---|---|---|---|---|---|---|---|---|
| Resource use and growth | Saving ratio in steep decline. Investment ratio also falling, down to 12% in 1989. Per capita incomes static. | Investment ratio *c.* 20%. Saving record improved and also high. Per capita incomes growing moderately. | Absorption/GDP declining; investment down to 10% in 1989. Per capita incomes declining. | Domestic saving increased and investment ratio stable. Some increase in per capita incomes. | Fixed investment and saving remain depressed. Growth revived with per capita incomes rising at 2–3%. | Investment fluctuating *c.* 20%. Growth remains slow and per capita incomes lower than early 1980s. | Saving and investment both down. Growth fluctuating; per capita incomes static. | Savings ratio improved; growth rate up and per capita incomes rising. |
| Government budget | Poor data. Current surplus and capital spending both declining. | Improved. Surpluses on current a/c, small overall deficit, but modest capital spending. | Revenue ratio falling, expenditure ratio rising. Deficits increasing. | Expenditure ratios cut in final years and also overall deficit. | Revenues revived and current a/c strengthened. But overall deficit remains substantial. | Budget balance improving slightly. | Some reduction in deficits, based on expenditure cuts. | Revenue ratio raised and budget balances improved. |
| Overall appraisal | Mixed record. No clear improvement. | An overall improvement. | Unambiguous deterioration. End-of-period economic crisis. | Clear overall improvement but fragile and data unreliable. | Clear overall improvement but mixed record and aid-dependent. | Difficult to read. Overall position still fragile. | A mixed, generally unsatisfactory record. | Impressive all-round improvement. |

*Note*: [a] RXR = real exchange rate.

## *The social and political effects of stabilisation*

### Distributional consequences

We suggested in the section on methodological issues that the case-study approach is better suited than cross-sectional analyses to the evaluation of programme impact on poverty and inequality. While this is true in principle, it has to be admitted that the literature surveyed here does not offer rich pickings; there is therefore a limited amount we can say.

One of the few reasonably firm things that the section on existing literature could report concerning distributional effects was Pastor's (1987) finding for Latin American countries that Fund programmes were strongly associated with declines in labour's share in the functional distribution of income. Some of our country studies are consistent with this (and none contradicts it). In the Gambia, for example, a wage freeze combined with rather rapid inflation to reduce substantially the real value of public sector earnings; the numbers employed by the state were also reduced. The wage labour force suffered in the Philippines too, where programme adoption was associated with 40–50 per cent falls in real wages and declines in employment in all sectors except services. There were also employment losses in Ghana's public sector, although there the evidence on real wages is more mixed.

Ghana, in fact, illustrates the difficulty of arriving at firm judgements about distributional effects even in detailed country studies, for the indications are that the effects were quite complex and there is little hard data to bring to bear on the subject. It is also a fine judgement, particularly in this case, as to whether adverse changes should be attributed to the programmes or to the long period of mismanagement which preceded them. Two groups of Ghanaian losers can be identified with some confidence: those who lost their jobs through retrenchments and those privileged persons who previously were able to enjoy the large scarcity rents bestowed by access to import licences and which the programme largely eliminated by drastic devaluations and other liberalising measures.

One major group of gainers is also clear: those whose incomes were derived from exporting, including large numbers of small-holder cocoa farmers, for the profitability of exporting was greatly increased by the programmes. The position of other important groups is less clear. In particular, it is uncertain whether farmers growing food crops, and their dependants (who probably comprise a majority of the poor), experienced net gains after 1983. The position of the large numbers working in private sector service activities is similarly obscure. The resumption of economic growth, reported earlier, was, of course, essential for the capability of the economy to reduce poverty but, against this, there was widespread concern that education, health and other social services, which were near collapse by the early 1980s, remained in poor condition.

Deteriorating social conditions were also a complaint about the record of Jamaica in the first half of the 1980s. Devaluations raised food prices in a country heavily dependent on food imports, and the purchasing power of low-income households declined. There were real cuts in social service expenditures, particularly in health, large reductions in food subsidies and evidence of growing undernourishment among schoolchildren. The government sought to reduce the social impact of its subsidy cuts with a targeted food aid programme, which included provisions for the feeding of schoolchildren, food supplementation through health clinics, and the distribution of food stamps targeted on the elderly and the very poor, although some of the latter slipped through the net.

The Costa Rican government was another which sought to protect the poor from the effects of its stabilisation programmes. Wage adjustment favoured low-paid workers; a programme was introduced which distributed food parcels to about 40,000 needy households; and (against some IMF opposition) subsidised credit was provided to smaller farmers and cattle ranchers. The Ghanaian government similarly introduced a 'programme to mitigate the social costs of adjustment', although it was not part of the original programme planning and was slow to have much impact.

What can we learn from these scraps of information? The most they do is to confirm (a) that stabilisation programmes are liable to have appreciable effects on the distribution of income but that these are apt to be rather complex and to vary from one situation to another; (b) that groups of the poor can indeed be among the losers, with the urban working class particularly at risk; (c) that governments adopting Fund programmes are none the less free to adopt measures to protect vulnerable groups, although there may be hard negotiations with the Fund over measures which might create large claims on public revenues. The priorities of the government in power, rather than those of the IMF, are probably the principal determinant of the ways in which programmes impinge upon the poor.

## Political effects

What about the political consequences of IMF stabilisation programmes? There are more instances in our sample where the Fund's support helped maintain governments in office (in that sense promoting political stability) than of it being destabilising.

We have already alluded to the indirectly political role of the Fund in the Gambia, where its support helped to strengthen the government's position and reduce the threat of involuntary incorporation into Senegal. Relations with Sudan during the Nimeiry period (i.e. until 1985) provide another, more dubious, example of the Fund as stabiliser, where there was a well-documented (Brown, 1990: Chapter 4) history of approving new programmes despite persistent delinquency in the execution of past programmes and the government's evident lack of commitment to macroeconomic management. Brown sees the Fund's support as pivotal, with its 'seal of approval' legitimising foreign assistance in the forms of debt relief, commodity aid and direct BoP support, without which it is doubtful whether the Nimeiry regime could have survived. Whether such political stabilisation was desirable is a different matter, however. Brown suggests that the financial support it received may have permitted the Nimeiry government actually to postpone economic reforms, and that by the time it was overthrown the country was not only in substantial debt but, above all, less resilient to external shocks.

The Philippines during the Marcos period provides a scarcely less notorious example. Here too there was a succession of weakly executed programmes which had similar effects to those just described for Sudan. This story has a twist to its tail, however, for by the time of the negotiations for a December 1984 stand-by it has been suggested that the IMF had had enough and that it 'changed its posture from that of a doting parent to that of a vengeful god' (Montes, 1987: 15). Montes goes on to suggest that the toughness of that programme and the desperation of the Marcos administration by then to adhere to it, even though that necessitated unpopular measures, increased the government's growing isolation from the business community and urban population, contributing to its downfall in 1986. We shall be returning to this and the Sudanese case later.

There are only two countries in our sample where relationships with the Fund impinged importantly on political stability. One of these was an EFF programme negotiated with the government of the Dominican Republic in 1983, which was abandoned the following year after the Fund pushed the government to take more stringent measures following weak initial implementation, against a background of arrests, riots and deaths. There were IMF-related strikes and protests during 1985 also, when a stand-by programme was negotiated, but by this time the government was more skilful in its political management and the credit was fully utilised.

Jamaica is our second destabilisation case. Here the IMF has occasionally interacted with the island's political processes in a rather intimate way. The story begins in the 1970s, when it seems likely that, in first adopting a relatively 'soft' negotiating stance and then hardening it, the IMF initially supported and then undermined the PNP government of Michael Manley (Sharpley, 1984). This led in 1980 to an 'IMF election' which saw the defeat of the PNP and the accession of the Jamaica Labour Party under Edward Seaga. There followed a honeymoon of about three years when the Fund showed considerable (some said surprising) flexibility despite substantial slippages in programme implementation. Following an (uncontested) election in 1983 and in the face of a considerable economic crisis, the Fund's stance hardened, however, and relations with the Seaga administration became more strained. Further slippage in programme implementation did not, however, prevent a national strike in June 1985 that was overtly linked to programme negotiations, and extensive unrest in the following year. It appears, however, that relations with the Fund did not feature in a major way in the 1989 election campaign and when the PNP was re-elected it 'reconfirmed its commitment to the medium-term adjustment strategy and announced policies to rejoin the adjustment path' (Robinson and Schmitz, 1989: 33) – a commitment which, however, did not prevent the existing stand-by programme from breaking down in March 1990.

It is difficult to derive radical conclusions from these cases. The Fund's role is inevitably political, for it is seeking to determine or influence policies which directly affect the well-being of virtually the entire population, including highly organised special-interest groups. The extent of its public profile in a country, the attitudes of the people towards it, the negotiating stances it takes up, the degree of flexibility it is willing to display and its decisions whether or not to approve a programme can make a large difference to a government's ability to move the economy forward, and to the programme's perceived legitimacy. Sometimes these factors will work so that the Fund has a stabilising effect, sometimes they will cut the other way. In line with Sidell's (1988) findings for the 1970s, there is little evidence that the net effects are systematically destabilising. It could well be the other way round; the Fund is, after all, helping governments cope with economic crises and providing financial support. However, there is a more disquieting side to this, as we shall see shortly.

## Stabilisation versus structural adjustment?

It is sometimes suggested that the Fund's short time horizon and its concentration on demand management are not merely inappropriate for those developing countries whose BoP difficulties stem more from structural weaknesses, but actually get in the way of structural adjustment. Our country evidence does not support this as a general proposition, but it does point to areas of tension.

We should recall first the broadly positive outcome of the above comparison of the start and end situations of our 'frequent user' group of countries (Table 3.6) – an analysis which takes a longer-term horizon – and point out that a high proportion of the countries analysed also had World Bank-supported structural adjustment programmes (SAPs) in operation during at least some of the period.[39] This is at least suggestive of an absence of any substantial conflict between the two programme types.

Two of our country studies provide documented examples of strong degrees of complementarity between stabilisation and adjustment programmes. In the Gambia, negotiation and implementation of both types of programme went in tandem and there was close collaboration between the Fund and the Bank. In consequence, the two thrusts were mutually reinforcing, with necessary action in the fiscal and monetary areas providing a more stable framework within which to improve and rationalise price incentives and in other ways strengthen the 'real' economy.

Ghana provides a similar example, and another case where the two lending institutions, despite hiccups, are evaluated as having worked well together and with the Ghanaian authorities. Indeed, Ghana's situation in 1982 was clearly one where it was impossible to contemplate any adequate action to restore the crumbling infrastructure and productive system unless the macroeconomy was also brought into better balance. Having achieved a fair degree of stabilisation, it was subsequently possible to make progress, albeit limited, in strengthening the productive structure.

Several of the other studies also attest to such complementarity, and to reasonable – and generally improving – co-operation between the Fund and the Bank. This is true, for example, of the literature on Jamaica, where the Bank's own assessment of the reasons for the failure of its first two SAPs identified the breakdown of the IMF stabilisation efforts as the most important single explanation. It also spoke of intensifying co-operation between the two institutions during the second half of the decade. Thus Harrigan on Jamaica (1991: 356–7):

> The post-1985 period . . . indicates that stabilisation and adjustment with growth are compatible provided the burden of the contractionary demand policies falls upon consumption and on *selective* areas of public expenditure. The clear implication is that although liberalisation, incentive structures and international competitiveness, which receive the focus of the Bank's policy reform thrust, are necessary to the generation of export-led growth, they are not sufficient. Equally important are the areas of fiscal, monetary and exchange rate policies, generally characterised as the province of the Fund's short-term demand management programmes.

Fund–Bank co-operation is also recorded as having being close in Côte d'Ivoire and Pakistan.

There do remain areas of difficulty in achieving consistency between the demand-management and supply-side tasks, however. The largest of these is with investment, particularly public-sector investment, for Harrigan's stipulation that the main burden should fall upon consumption is only exceptionally observed. We have earlier shown that investment was depressed in a large proportion of our countries and that IMF (and World Bank) programmes are generally associated with reduced investment levels. The central point here, of course, is that successful adjustment of productive and institutional structures necessitates large new investments; approaches to the control of absorption which reduce investment are apt therefore to frustrate such adaptation.

Of course, large reductions in investment are not normally an intended result. We showed in our discussion of the existing literature that targeted investment levels were well above realised results and that this was an area in which Fund programmes were least successful in achieving intended outcomes. Within the public sector, at least, the problem arises from the political and other difficulties of cutting back sufficiently on the recurrent (consumption) budget and the temptation, therefore, to come down first and heaviest on capital spending (as in the 'shock' phase of Mexico's 1983–5 EFF programme, when there was a 35 per cent cut in real public investment in a single year). Whether the Fund could have done more to guard against this is an important but unanswered question. What is clear, however, is that it could only have done so at the risk of prompting more allegations of interference with the home government's priorities and of reducing public-sector employment yet further.

Another potentially important source of stress between stabilisation and adjustment has to do with the supply of imports. One reason why demand management is seen by the IMF as so central to improving the BoP is in order to reduce the import bill. Imports, however, are important, often indispensable, for the supply of capital and intermediate goods, without which it is impossible to raise resource utilisation and restructure the productive system. There were symptoms of such an import strangulation in only a few of our cases, however.

This has already been mentioned in the case of Malawi, where there was a 35 per cent fall in import volumes in the 1980–8 period, implying even larger falls relative to GDP and to population. Much of the improvement in Jamaica's current account from 1983–5 is similarly attributed to reductions in imports, resulting from a demand squeeze. The Philippines under Marcos underwent a severe squeeze in the first half of the 1980s, with import volumes declining by around a third during 1980–5. To what extent either of these experiences could correctly be attributed to the Fund's programmes is unclear, however.

Moreover, we should set against these the Ghanaian programme which incorporated provisions for increased quantities of imports. The remaining countries, rather to our surprise, provided little evidence of major import cuts of the type that could undermine structural measures, and Bangladesh (from 1981–2) had to cut back on imports consequent upon the *abandonment* of a Fund programme. Nor does the literature under review record major difficulties between the Fund and the Bank on this matter. We can further recall that, while Table 3.5 did show IMF programmes to be associated with import reductions, these cuts were not large and had generally modest levels of significance.

Finally, we should recall from the same source the evidence of a strong association between adoption of Fund programmes and sustained real depreciation of the exchange rate – a result borne out for most, although not all, the countries surveyed here. The relevance of this is that the exchange rate has a strong influence on the allocation of resources between tradables and non-tradables. As such, IMF-initiated devaluations must be regarded as strongly supportive – often to the point of being a necessary condition – of supply-side programmes.

## Influences on programme effectiveness

Our next task is to explore what light the country literature can throw on the determinants of programme effectiveness. We begin by looking at the influence of exogenous shocks.

### *Exogenous shocks*

Our starting point here is to recall the result from Table 3.5 which showed that our sample of 17 countries experienced deteriorating terms of trade in the 1980s, but also to recall that the extent of this was greater amongst those countries whose programmes were concluded prematurely than for those whose programmes lasted their intended course. This suggests that deteriorating trading conditions are an important influence on programme effectiveness.

Somalia is a country whose history is consistent with this hypothesis, for it is noticeable that programme breakdowns in 1981 and 1986 were both accompanied by substantial declines in the terms-of-trade index, whereas it was stable during the successfully completed programmes of 1981–2 and 1982–4. Of course, there were other important influences on the outcomes of these programmes. This is also true of Tanzania's short-lived 1980 programme, but to which an unexpected collapse of tobacco and tea prices, plus poor weather affecting cotton and maize output, also made a contribution. Exogenous shocks also affected outcomes in Côte d'Ivoire, where a drought in 1983 was a second-order contributor to the breakdown of an EFF programme.

Malawi provides perhaps the strongest illustration of external factors undermining programme effectiveness. The problem there was not confined to adverse movements in world prices, however, being aggravated by the costs imposed by the destabilisation in southern Africa during the 1980s. In Malawi's case, the most serious aspect of this was the civil war in neighbouring Mozambique. This severed rail links to the coast, greatly increasing transport costs, and caused a substantial influx of refugees, as well as necessitating levels of military expenditure Malawi could ill afford.

A number of examples in our sample suggest a link between external shocks and fiscal targets, and consequent success or failure in meeting performance criteria. The 1980/81 EFF in Morocco is an example of a programme undone by an adverse shock. It was suspended following a drought and an increase in food imports, with a higher food-subsidy burden contributing to a larger-than-expected fiscal deficit and the breach of domestic credit ceilings. Similarly, credit ceilings under the 1979 stand-by in Malawi were breached following harvest failure and the need to import food on commercial terms. Another example is provided by the 1981 EFF in Côte d'Ivoire which was (partly) disrupted by declines in international prices, leading to tax revenue shortfalls, with the government committed to fixed domestic procurement prices.

On the other hand, favourable conditions do not necessarily ease the attainment of performance criteria. In Malawi, maize prices were raised in a bid to improve food self-sufficiency. The subsequent supply response and favourable weather increased the budgetary cost of procurement to the extent of breaching the 1983 EFF credit ceilings. Similarly in Pakistan and Bangladesh: their respective 1980 EFFs were suspended following bumper harvests, commodity operations to build buffer stocks and the subsequent breach of credit ceilings. In Bangladesh's case, the agreement was suspended despite compliance with other performance criteria.

By contrast, the increase in international prices for Ivorian coffee and cocoa exports during the 1984 stand-by agreement facilitated the attainment of fiscal targets. Similarly in Gambia, failure to pass on falling international oil prices to domestic consumers boosted government revenues and facilitated compliance with the 1986 stand-by agreement.

To keep things in perspective, none of the programmes studied by our country literature provides an example where exogenous forces palpably dominated outcomes, making it impossible for objectives to be realised. In at least two cases – the Gambia and Morocco – programmes were actually assisted by favourable turns in the terms of trade. And the Ghanaian case shows that major progress can be made even in the face of initial problems of the greatest severity and strongly adverse international price movements, although Ghana's programmes admittedly received abnormally large amounts of supporting finance.

Against this, we must recall the fragility of many of the achievements of the 1980s and the vulnerability to external forces of a high proportion of our 17 countries. Thus, even in the relatively prosperous economy of Jamaica authors emphasise the limitations of what can be achieved by domestic policy actions, given the strength of the outside forces working upon it. Much the same could be written of the majority of our sample.

### *Financial flows and catalytic effects*

Is there any evidence that failures to induce expected new inflows of capital put programmes at risk through under-funding? It is certainly easier to pick out examples from our sample which point in this direction than cases where there was a demonstrable catalytic effect. In fact, only two countries fall clearly into the latter category. We have just mentioned the favourable experiences of Ghana, which from the mid-1980s was able to attract large amounts of capital from the World Bank and other aid donors. However, we have also mentioned that even there overall investment levels remained drastically low and there was little sign of any revival in private capital inflows.

The Gambia is the second of the countries which was able to reverse pre-programme donor disillusionment, so that donor BoP support shot up from the equivalent of 5 per cent of GDP in 1985/6 to 44 per cent in 1986/7. Much of this, moreover, was in grants, permitting large improvements in both BoP and budgetary current account out-turns, and a corresponding easing of inflationary pressures. Of course, in attracting such large levels of support Gambia is a classic beneficiary of the small-country effect,[40] leaving doubts about the viability of progress should aid fall back to 'normal' levels. And here too it does not appear that there had been any major revival of private investment.

The absence of any measurable net catalytic effect is evident in a much larger number of our countries: Bangladesh, Costa Rica, Côte d'Ivoire, Malawi, Mexico, Pakistan and Somalia. This is particularly noteworthy since for many of our sample agreement with the IMF was linked with debt reschedulings (which, however, were in this period typically short term and ungenerous). It seems clear that often the Fund credits were, in effect, permitting financial outflows to commercial banks and others – and, in some cases, capital flight. Hence, in Jamaica increased inflows of public finance tended to be offset by declining private inflows, although we should bear in mind the generally weak record of programme implementation in that country.

It is, of course, difficult to draw a close connection between the weak response of the BoP capital account and programme effectiveness, and we have already noted the lack of strong evidence of failures resulting from import strangulation. However, we have also dwelt upon the strongly adverse longer-term implications of the depressed investment levels found in most of our countries – levels that would have been higher had a catalytic effect been more in evidence. At least as regards private investment, the Fund is probably unrealistic to expect any strong response in normal conditions, for our evidence (and most of that surveyed in the discussion of existing literature) suggests that the economic results obtained from Fund programmes are neither substantial nor dependable enough to provide a firm basis upon which the confidence of private investors is likely to be restored. There is also the possibility that programme negotiations signal the existence of financial difficulties of which potential private investors were previously unaware. We are thus left with a conundrum: do programmes fail because foreign investment does not respond, or does foreign investment stay away because programmes fail? Or, indeed, simply because they exist?

### *The difficulties of controlling the budget*

Earlier sections have drawn attention to the difficulties which Fund programmes have in trying to improve the fiscal balance. The cases surveyed here similarly had varied and often negative fiscal experiences. Thus, in Bangladesh's 1980–83 EFF it was in the fiscal area that the only major slippages in programme execution occurred, with an increase in current expenditures more than twice as fast as that written into the programme. In Côte d'Ivoire too it was found impossible to contain recurrent expenditures as planned in the 1981–4 EFF, due largely to unplanned increases in the civil service wage bill and in external interest payments. In Morocco's 1986 stand-by it was similarly only in the fiscal area that there was substantial slippage.

In Mexico there were also large slippages, as is shown by the following comparisons of programme and actual values for the public sector borrowing requirement (expressed as percentages of GDP) (Ros and Lustig, 1987: Table 3); 'target' figures are the original IMF projections:

| | *Target* | *Actual* |
|---|---|---|
| 1984 | 5.5 | 8.7 |
| 1985 | 3.5 | 9.9 |

There was a smaller number of opposite cases, where Fund programmes were associated with major improvements in the fiscal situation. We have already mentioned Ghana, where it was possible to transform the current budget from a large deficit to a modest surplus while also raising expenditure ratios, due to the positive impact of the programme on the revenue base. Dominica provides another example, where tax measures under the 1981–4 EFF substantially raised the revenue base and this combined with an effective restraint of expenditures to bring a much-strengthened budgetary outcome. In the Gambia there was a combination of strengthened expenditure controls (including reduction in the size of the civil service) and revenue-increasing measures, although these appear to have brought a rather modest overall improvement in the budgetary balance.

Encouraging though they are, these positive examples are the exception rather than the rule, and we were left with the impression that the frequently limited impact of programmes on the public finances (other than capital expenditures, which is arguably among the least desirable area for cuts) is an important reason why programmes are often unable to achieve their objectives. In turn, some of the other sources of difficulty discussed in this section tend to aggravate the fiscal problem. Trade shocks can reduce revenues from import and export duties and make budgetary planning more difficult. Rises in world interest rates can increase the demands of interest payments on the exchequer. Bad weather can both reduce revenues and create additional spending needs. Shortages of supporting capital can tighten the budgetary situation and add to the temptations of large-scale deficit financing. Geo-political pressures in favour of a developing country government can weaken its resolve to confront hard political choices in order to strengthen the budget. Our case materials include examples under each of these headings.

### *Government–Fund relations*

There have been a number of well-publicised instances in which relationships between the home government and the IMF have been adversarial. In many more the Fund and others have nominated 'inadequate political will' as a key explanation of failed programmes. What light does our sample throw on Fund–government relations?

First, it contains more examples of fairly easy relationships, with substantial consensus on what measures should be taken to achieve stabilisation, than it does of the opposite situation, although it is in the nature of things that negotiations which fail to result in agreement may well be neglected in the literature. In Malawi, for example, there was substantial congruence of outlook between visiting Fund missions and the government, and there is evidence of good levels of co-operation (also encompassing the World Bank). The same is recorded for Costa Rica in the early 1980s, when the 1982–3 stand-by programme was evidently based largely on the Monge administration's own assessment of what needed to be done. There was a similar lack of tension in the Gambia, once the government had decided on the necessity for a change of tack, although we may surmise that in this country programme design was dominated by Fund staff members and other foreign advisers. Other country episodes marked by IMF-government congruence include Pakistan in 1980, Jamaica from 1980–81 (although relations subsequently turned sour), Mexico in 1983, and Ghana in the later 1980s.

However, our sample also throws up conflict cases. The negotiations leading to Tanzania's abortive 1980–82 stand-by provide an almost pure form of this (although things improved later). President Nyerere's team brought to the negotiations a socialistic set of priorities which emphasised distributional over efficiency concerns, showed little faith in the efficacy of market incentives, and was inclined to see the IMF as a hostile agency of international capitalism. With the Fund mission apparently under instructions to take a tough line, negotiations were protracted, sometimes acrimonious, centring around the desirable extent of devaluation. Unsurprisingly, the resulting programme broke down within three months.

We referred earlier to a stiffening in IMF attitudes towards the Marcos administration in the Philippines in 1984. The negotiations for the stand-by that was eventually approved late that year were long and difficult, complicated by National Assembly elections in May, in the run-up to which the government ignored the economic situation. However, this was scarcely the result of a clash of competing political philosophies. Negotiations for Mexico's 1986 stand-by are also reported to have been difficult, even though the government had already demonstrated a rather conservative approach to macroeconomic management.

Jamaica is a country in which, as already described, the ease of the Fund's relationships, even with the same administration, has gone through marked cycles and where the Fund and the country's leaders have generally found it difficult to reconcile the political and institutional imperatives under which they respectively act. The case of Ghana, by contrast, appears to be one of a developing relationship. It seems apparent that the Rawlings government accepted the volte-face in economic policies represented by the 1983 programme only because it could see no alternative, and that Fund (and World Bank) staff dominated the preparation of that programme. By the end of the decade there was much more of a genuine consensus and the relationship was a more equal one.

What our case materials also demonstrate, however, is that *congruence provides no guarantee of implementation*. There are several documented cases of apparently consensual programmes which were poorly implemented. Thus, Brazil's 1983 EFF followed an orthodox stabilisation programme initiated by the government in late 1980, with the terms of the programme quickly agreed. But, while some of its provisions were undertaken, political and bureaucratic resistances were allowed to prevent the improved fiscal discipline which stabilisation demanded. Côte d'Ivoire provides another illustration. In its general conservatism the government of Houphuët Boigny was close in its attitudes to that of the IMF.

None the less, its record in programme implementation was indifferent, particularly in the area of price incentives. We have already covered the case of Jamaica's 1981–4 EFF, weakly implemented despite the coming to power of a government apparently committed to macroeconomic rectitude and considerable softness on the part of the IMF in negotiating the package.

It is, then, easy to exaggerate the frequency with which programmes are dictated by the Fund to a reluctant but desperate government. That situation is not typical. But even though those programmes which *are* dictated are among the least likely to be executed or sustained, agreement between the two parties provides no assurance of good implementation. In many cases, programmes are inaugurated in crisis conditions and to implement them requires skills in political management which are sometimes beyond the competence of the government, however well-meaning it may be.

### *IMF rigidity*

One of the most common complaints about the Fund, particularly when governments are seeking to stabilise against the background of economic and/or political turbulence, is that it is too rigid in its negotiating stances, too uniform in its approach to programme design. Our case materials provide only limited information on this, but there is a remarkable absence of complaints about inflexibility.

Matin's (1986, 1990) writings on Bangladesh's programmes are an exception, arguing the need for a more judgemental approach by the IMF to decisions about continuing access to a credit, rather than a perceived mechanistic approach based on observance of quantified performance criteria. Our earlier discussion of the fiscal effects of exogenous shocks also suggested that the Fund was on occasion too rigid to adjust performance criteria in the light of changed circumstances, leading to programme breakdowns. Our materials also include many cases in which programmes were suspended or abandoned, and some of these breakdowns may have been the result of Fund rigidity. However, if we confine ourselves to the literature under review, there are more examples of apparent flexibility than of the opposite.

We should, however, make a distinction between that type of flexibility which leads the Fund to take up a 'soft' position during programme negotiations and flexibility in their execution. There are quite a lot of examples of the former, only one or two of the latter. The Fund could be described as having taken up relatively soft positions – for reasons we return to shortly – in the Dominican Republic (1983); Jamaica (1981); Philippines (1979–83); Sudan (1979–84); and Mexico (1986, 1989). The latter country is of special interest because IMF flexibility was, as it were, formalised in two programmes that contained unique features. Thus, the 1986 stand-by contained provisions that varied the terms of the programme depending on external trading and other conditions, with both automatic modification of performance criteria and of the size of the credit according to the value of petroleum export earnings and the growth of the economy. These terms caused excitement among other developing countries at the time as possibly signalling a more universal IMF move towards programme flexibility, although it subsequently became clear that Mexico was to be treated as a special case. Similarly, its 1989 EFF was regarded as a landmark in explicitly recognising the effects of the debt overhang and adopting a medium-term growth framework.

The one documented example we have on Fund flexibility in programme implementation relates to the Gambia (although there were almost certainly other instances which our literature does no more than hint at). In this example, credit to government exceeded the programme ceiling for technical reasons associated with the government's premature distribution of STABEX (aid) receipts. The Fund staff apparently agreed that the transfer made good economic sense and access to the credit was not jeopardised.

### *Fund-over optimism*

The section on existing literature (see pp. 45–6) refers to an apparently chronic tendency for Fund missions to be over-optimistic in their programme negotiations. This can relate to diagnosis of the initial situation, to the ease with which their introduction of policy changes is envisaged, and to the size and speed of expected economic responses to these changes. Our cases contain a number of illustrations. The strongest relates to Jamaica, where Harrigan (1991) points to excessively sanguine forecasts in 1983 by Fund staff of prospects for the local bauxite industry and for private capital inflows as a basis for performance tests which then proved beyond the reach of government. She suggests, however, that these errors may not have been so much a matter of poor judgement as a necessary artifice to enable a programme to be cobbled together.

Similar over-optimism about external developments is recorded for Côte d'Ivoire in respect of the state of world demand, and also about prospects for the development of the domestic petroleum industry. Mexico's 1983 programme was influenced, possibly undone, by mission optimism about the possibilities of combining renewed economic growth with import restraint. The Dominican Republic's 1985–6 stand-by is recorded as not having achieved the expected results because the Fund was unrealistic about the speed and extent to which imports and exports would respond to a devaluation. Overall, our sample underscores both the existence of a bias towards over-optimism and its potentially serious consequences for the feasibility of some programmes.

### *Geopolitical interference*

The Fund's official position is that it applies uniformity of treatment in country programmes, which is defined to mean that, for any given degree of need, the effort of economic adjustment should be broadly equivalent among members. In practice, it is sometimes prevented from applying this principle by the lobbying of major shareholder (i.e. industrial country) governments to secure easy terms for favoured borrowing countries, in pursuit of foreign policy or security goals. Our sample provides plentiful examples of such favouritism. Moreover, it is this phenomenon which explains virtually all the examples of IMF flexibility or 'softness' in negotiating programmes cited above.

We have already cited Brown's (1990) study of Sudan, documenting a US-inspired policy of Fund support to a regime which had little interest in macro stabilisation. We have similarly described the position of the Marcos government in the Philippines prior to 1984, and of the Seaga administration in Jamaica after its election in 1980. To these we could add, although slightly more speculatively, Côte d'Ivoire, which enjoyed a succession of programmes notwithstanding considerable slippages in implementation, aided by lobbying by the French government. Although it scarcely commands great geopolitical importance, even the Gambia appears to have benefited – in its case from international concerns about the threat of annexation by Senegal. It is also highly probable that pressures from the US and other creditor governments explain the uniquely favourable programmes Mexico was able to negotiate.

Thus, of our 17 countries, at least a third have secured favourable programmes because of special relationships with major shareholder countries. Indeed, a third is probably an underestimate because our sample almost certainly includes other examples which were not documented but which are widely regarded as having been the beneficiaries of this type of influence. Brazil is a case in point, as is Pakistan in our period. Other possible candidates, for which, however, we are unable to offer concrete evidence, include the Dominican Republic, Malawi and Morocco. Some would add the negative example of Tanzania in the early 1980s, as a left-leaning government faced with an IMF mission instructed to take a tough line.

Amongst those who know the Fund well there are disagreements about the extent and effects of political interference in country programme decisions. Because a major shareholder government takes an active interest in the terms of a programme, this does not necessarily mean the Fund staff are overruled. Sometimes they are successful in resisting such pressures. Also, we suspect that the incidence of serious levels of interference is less overall than that revealed by our sample. None the less, it is undoubtedly a real problem and, for our sample at least, *it provides among the strongest explanations for ineffective programmes*: in Sudan, the Philippines, Côte d'Ivoire, Jamaica, possibly other sample countries. In its more extreme forms it amounts to an unconditional provision of finance for governments with proven records of macroeconomic mismanagement. As such, it contradicts all that the IMF is supposed to stand for, and undermines its legitimacy in other countries whose governments resent the more favourable treatment received by others and which are tempted to try to secure by stealth an equality of treatment they are unable to negotiate formally. In this, the Fund itself is largely a victim, except to the extent that its management could be more resolute in resisting the political pressures to which it is subjected. It is the governments of major countries which are chiefly at fault, seeking simultaneously to promote contradictory financial and foreign policy objectives.

## CONCLUSIONS AND SYNTHESIS OF FINDINGS

The title of this chapter asks how much it is possible to know about the effects of IMF programmes. It is evident from the materials presented here that we can know – or anyway be reasonably confident about – quite a lot. However, the reader must by now be suffering from information overload, so in this final section we attempt to bring out the main points by first making some concluding remarks and then offering a synthesis of the main findings to emerge.

### Conclusions

One lesson that emerges clearly is that the world (and probably the institution itself) overrates what the Fund can realistically hope to achieve, and the amount of leverage which it exerts over policies. Although it exerts important influence over the exchange rate, the evidence suggests limited impact upon other instruments central to its programme designs. Relatedly, we have shown there to be extensive slippages in programme execution, a problem which has probably worsened in recent years.

At the same time, countries which adopt Fund programmes are often not only having to cope with severe initial difficulties, including natural disasters, which are only partly the result of past policy weaknesses, but also having to contend with adverse shocks during the course of the programmes. The superior power of such exogenous forces is one of the recurring themes of the empirical literature. To make matters worse, the Fund is not able to have the *predictable* catalytic effect on capital flows which is often claimed for it, and some programmes are associated with *diminished* net resource inflows. This particularly pertains to countries looking for increased private investment. Moreover, while we have shown that the average Fund credit is substantial relative to need, it is a good deal less so in the case of programmes which subsequently break down, consistent with the criticism that programmes often break down because they are under-funded.

The point is this: in the face of often adverse circumstances programmes cannot expect to bring substantial, lasting improvements unless they can change key policies in a decisive way. And even if they can do so, it remains an open question whether the power of the policy instruments is commensurate with the size of the problems addressed. While the evidence does suggest that the IMF is able to exert a considerable influence over price variables such as the exchange rate, interest rates and producer prices, this is not matched by commensurate influence over fiscal and monetary instruments, nor in the area of institutional reforms. Given the Fund's limited revealed leverage over these other variables, it is scarcely surprising if the macroeconomic results of its programmes fall well short not only of intentions but also of the degree of sustained improvement necessary to ensure credibility for government policies and to revive investor confidence.

Conditionality – the policy leverage bought by the provision of financial support – emerges as something of a toothless tiger. Such has also been the experience of the World Bank in connection with its structural adjustment programmes.[41] While simulations and other forms of evidence suggest that the types of policy change favoured by the Fund are, indeed, likely to strengthen the BoP, the evidence on both stand-by and ESAF programmes is that BoP results are not strongly affected by the extent of programme implementation. We can conjecture that what matters a good deal more is the extent to which the government itself is convinced of the need for prudent and credible macroeconomic management, in which the influence of the IMF may be secondary. The question arises as to whether conditionality is an effective modality for achieving policy changes – a question which points a dagger at the heart of the IMF's *modus operandi*.

At the same time, we understate the impact of the Fund by concentrating solely on the direct effects of its programmes. Over the last decade or more there has been a sea-change in developing country government attitudes towards the benefits of macroeconomic stability and fiscal-monetary rectitude. The basic thrust of what the Fund seeks to achieve in the area of economic management is far less contentious today than it was in the 1970s. Managing Director Camdessus likes to talk of a 'silent revolution' in government attitudes towards macro policies. For this conversion the IMF surely deserves a considerable portion of the credit.

However, the standing of the Fund in these countries is threatened by political lobbying of major shareholders in favour of (or against) particular borrowing countries, in promotion of their own foreign policy objectives. For our sample of countries, this was both common and subversive of what the IMF was trying to do.

The frequency with which it apparently occurred (together with the reluctance of major shareholders to expand the Fund's resources in line with needed levels of programme support) raises questions about the seriousness of industrial country commitment to the IMF's stabilisation mandate, although it may be that the ending of the Cold War will diminish this danger to the credibility of Fund programmes.

Perhaps the most central conclusion emerging from the evidence presented in this chapter is that it is difficult to understand the fierce controversies which have surrounded IMF programmes in developing countries, and the dramatically opposed positions that are sometimes taken up. Although, for different reasons, it suits both the Fund and its critics to assert that its programmes have large effects, for good or ill, the evidence presented here suggests otherwise. Programmes do often move the BoP some distance in the desired direction, not least by improving export performance, and programmes may have a more sustained influence than is often assumed. And, with the important exception of lowered investment (and, to a lesser extent, import) levels, it does not appear that programmes systematically result in large distributional, political or supply-side costs. In favourable circumstances the benefits doubtless well exceed the costs; in other cases the costs may be real and the benefits elusive. In the general case, however, it is all too easy to exaggerate the magnitude of both.

Overall, then, the message is that the Fund has limited ability to achieve its objectives and to assist deficit governments. On this we agree with Khan (1990: 222), a senior member of the Fund's research staff:

> one would be hard-pressed to extract from existing studies strong inferences about the effects of Fund programs on the principal macroeconomic targets.

That this should be the case is regrettable. Many developing countries badly need to strengthen their macroeconomies. The IMF is therefore seeking to meet a real need, the more so since it has moved to broaden the design of its programmes to comprehend supply-side and institutional factors. Unfortunately, our results, while allaying the critics' worst fears, leave plenty of reasons for believing that there is room for improvement. We take up this theme in Chapter 5.

Finally, it remains to offer a synthesis of the more firmly based findings reported in previous sections: a check-list of 'what we can know' about the effects of IMF programmes.

## Synthesis of findings

### *Sources of disequilibrium*

1. It appears that in a high proportion of cases a *combination* of exogenous shocks and domestic policy weaknesses is important in the events leading up to the adoption of a Fund programme. It is rare for either category of explanation to be materially absent, or for external factors to dominate. A frequent sequence is that exogenous shocks lay bare inadequacies of policy, hastening the need for change.

2. Our country studies suggest that natural disasters are a more frequent reason for programme adoption than is generally appreciated. They also emphasise the singular vulnerability of countries heavily dependent on primary product exports, not only because of adverse movements in the terms of trade but also due to mismanagement of short-term commodity booms. Slowness of policy response to an emerging problem is also a common ingredient of pre-programme crises.

### *Impact on target variables*

3. Overall, it appears that programmes are associated with a strengthening of countries' balance of payments. Before–after tests on the IMF's ability to improve on the BoP situations existing immediately prior to programme adoption provide conflicting evidence. Our tests for the 1980s suggest moderate improvements in export, current account and overall BoP performances in the short term, but substantial improvements thereafter (despite deteriorating terms of trade); earlier studies obtained weaker results. Before–after evidence on ESAF is similarly weak but positive. With–without and other approaches which seek to estimate the counterfactual situation are more encouraging to the Fund. The amount of such evidence is restricted and the results mixed, but they indicate that BoP performance is stronger as a result of programmes than it would otherwise have been.

4. The bulk of the evidence suggests little association between

Fund programmes and reduced inflation, with price-reducing and price-raising (e.g. from devaluations) effects tending to cancel out, although our 'successful completer' group showed some reduction.

5. Questions about programme potency are raised by evidence that BoP outcomes are not systematically correlated with programme implementation, either with respect to stand-bys or ESAF programmes. On the other hand, the results of with–without tests suggest that those which adopt IMF programmes do achieve better stabilisation records than those with similar initial problems which do not adopt programmes. A plausible explanation of the apparent contradiction between these before–after and with–without results is that governments reluctant to utilise the IMF are apt as a group to give lower priority to macroeconomic management, or simply to be incompetent in that area. If so, the governments that do use the Fund would tend to get better macro results even if the institution did not exist.

6. A substantial proportion of countries (including 11 of our sample of 17) have frequent recourse to Fund credits, contrary to its original objectives. From this it may be inferred that programmes are often unable to achieve sustained stabilisation. However, our study of the late 1980s situation of frequent-user countries suggests that a majority of them had somewhat stronger BoPs and lower inflation than ten years earlier, although the record on other performance indicators was variable. This finding was reinforced by statistical analysis of differences between the outcomes in cases where a Fund programme had previously been successfully completed and those where there had been no immediately prior programme, which found that the former group obtained superior BoP results.

7. Tests which analyse changes beyond the first year do not support the view that programme effects are short-lived. Indeed, impact is at its smallest in the twelve months immediately following adoption of a programme but larger in the following two years. It may be that the tendency in the past has been to evaluate Fund programmes over too brief a time, resulting in understatement of their effects. A corollary, however, is that governments which turn to the IMF for a quick fix are apt to be disappointed.

8. Delay or non-implementation of programme provisions is a large problem for the Fund. Only a fifth of ESAF programmes appear to have been completed on schedule and there has been a 50 per cent increase in stand-by non-completions since the mid-

1980s, although other information on tendencies in programme effectiveness shows no clear trend. At a minimum, it seems that implementation is not improving.

### *Programme costs*

9. Most tests indicate that adoption of an IMF programme is not associated with any significant loss of output or slowing of economic growth, although a minority suggest that there is a significant negative correlation. Simulations suggest that growth will be more favourable with programmes which incorporate investment-raising supply-side measures, and, consistent with this, SAF programmes appear to have done well in this regard. ESAFs have done less well, however, and Fund staff appraisals show that, rhetoric notwithstanding, growth is not treated as a major ESAF objective, constraining programme design. Overall, there is little evidence that programmes generally have substantial growth-stimulating or growth-retarding effects.

10. Any slow-down in growth may be the result of the substantial and sustained depression of investment rates, which tests rather firmly link with the adoption of Fund programmes. There is also evidence that some Fund programmes are linked to reductions in import volumes. This is also liable to worsen economic performance, although our country studies uncovered little evidence of severe 'import strangulation'.

11. Programme association with reduced investment is a serious source of difficulty in reconciling short-term BoP stabilisation with longer-term structural adjustment. This may also be true in specific cases of programme-related import cuts. In other respects, however, stabilisation via the IMF appears complementary to structural adjustment; and Fund–Bank co-operation seems to function quite well.

12. Available information on programme consequences for income distribution suggests (a) that stabilisation programmes are liable to have appreciable effects, but that these are apt to be rather complex and to vary from one country to another; (b) that groups of the poor may indeed be among the losers, with the urban working class particularly at risk; but (c) that some governments adopting Fund programmes have been able to adopt measures to protect vulnerable groups, although there may be hard negotiations with the Fund over measures which create large

claims on public revenues. The priorities of the government in power, rather than those of the IMF, are probably the principal determinant of the ways in which programmes impinge upon the poor.

13. Anecdotal evidence notwithstanding, there is no basis for the suggestion that Fund programmes result in systematic political destabilisation. IMF support may have the contrary tendency by supporting incumbent governments, but whether enhanced political continuity is actually desirable depends on regime and country circumstances.

## *Determinants of programme effectiveness*

### Exogenous determinants

14. There is evidence that both the gravity of the initial situation and the intrusion of unforeseen shocks are major reasons for programme failure, partly because of the vulnerability of many developing countries. Programme breakdown is associated with particularly adverse terms-of-trade experiences. On the other hand, our country survey found no example where exogenous factors dominated outcomes.

15. Programmes may fail because of inadequate supporting finance, and it is sometimes argued that Fund credits are too small to provide adequate support. The view that its credits are generally small is disputed for our country sample, which shows that the average credit, net of return flows, had a value equivalent to nearly a tenth of imports. However, the view that under-funding is a source of programme failure receives support (although of modest statistical significance) from evidence that programmes which are prematurely discontinued receive only about half the average amount of credit.

16. Relatedly, although the Fund intends that its programmes should have a catalytic effect on the inflow of foreign capital (including debt relief and repatriation of flight capital) and frequently claims such a result, revealed effects on the BoP capital account are complex. Although catalytic effects can be observed in specific cases, the evidence of systematic net additional inflows is mixed. There appears to be no general effect on private flows but there may be an inducement of additional public transfers. Some

Fund credits have been used to repay other creditors, and some programmes probably fail because of inadequate finance, although it is difficult to demonstrate this connection.

### Influences relating to the IMF and its programmes

17. Programmes have only a limited impact on several key macroeconomic policy variables. The evidence is no better than mixed even on the core programme component of domestic credit, for which there is little statistically significant evidence of effective restraint. Information on fiscal impact suggests that, while there is a tendency for budget deficits to be reduced, this effect is weak and variable, and there is much slippage in the implementation of programmes' fiscal provisions. There is also evidence of much slippage in the implementation of public enterprise reforms, privatisations and other institutional changes.

18. Real exchange-rate depreciations are an important exception to the Fund's generally muffled influence on policy. Our findings suggest not only that programmes are associated with a substantial and significant real depreciation but that this is sustained and deepened in post-programme years, i.e. is not neutralised by inflation. Fund programmes probably exert considerable influence over other price variables, including interest rates and public enterprise prices. Overall, however, the Fund appears unable to change policies as much as it would like to – and needs to do for programme success.

19. This is no doubt partly because there is much foot-dragging by governments in the execution of programme provisions. Half of stand-by programmes break down during their intended lives, usually as a result of non-compliance, and there is other substantial evidence of incomplete programme execution. As already noted, the economic consequences of this do not appear large, however.

20. Where they exist, adversarial relationships between governments and the Fund are likely to contribute to non-implementation of policy conditions, but such relationships appear exceptional. In most of our country studies the relationship was a workmanlike one, no doubt with negotiating disagreements on specifics but little apparent clash in general approach.

21. Good working relations provide no guarantee of programme execution, however. Our studies found several documen-

ted cases of apparently consensual programmes which were poorly implemented, with governments unwilling or unable to override local political opposition to programme provisions. Programmes are often inaugurated in crisis conditions and require for their enforcement political resources which the governments in question may not possess.

22. Programmes have a consistent tendency towards over-optimism in their BoP and inflation targets, with several studies showing only about a 50 per cent success rate. In addition to technical and data problems, reasons for this include limited leverage over government policies and pressures on Fund missions to agree programmes which they privately regard as unrealistic. Particular unrealism is experienced in programme targets for GDP growth and investment. Such over-optimism has serious consequences for programme feasibility and adequacy of supporting finance.

23. Although complaints about IMF rigidity in programme negotiations are common in public discussions, our country survey found limited documentation of rigidity and substantial evidence of flexibility in negotiations, although not in programme implementation. On implementation, there were a number of cases where the Fund had apparently insisted on maintaining unchanged performance criteria in the face of substantial external shocks beyond the control of government.

24. During the period studied, much of the flexibility in negotiation was apparently the result of political lobbying on the part of major shareholder governments for favourable treatment of friendly governments. This affected a considerable proportion of the 17 countries studied. The Fund was hence prevented from observing its principle of uniformity of treatment. In its worst forms, such political interference forced the Fund to provide essentially unconditional finance to governments with proven records of economic mismanagement. This undermined its credibility, and was among the most important reasons for programme ineffectiveness among the countries studied. With the ending of the Cold War, the importance of this factor may have diminished, however.

# 4

# ISSUES IN THE DESIGN OF IMF PROGRAMMES

Earlier chapters have examined the extent of continuity and change in the content of the IMF's programmes and evidence on their effects. We turn now to survey some unresolved issues concerning the design of these programmes. We start with a brief exposition of the basic policy model used by the Fund. The bulk of the chapter is then taken up with discussion of issues arising, concerning: the uses and limitations of the model; the policy instruments and performance criteria employed in the programmes; the problem of programme inflexibility; and the difficulties the Fund has in handling the political aspects of stabilisation policy.

## THE FINANCIAL PROGRAMMING MODEL

### The basic model

The analytical core of IMF programmes is what it calls its 'financial programming' model, although we shall see shortly that this provides only a starting point.[1] This is also known as the Polak model, after the Fund's former Director of Research who first presented this form of monetary model of the balance of payments (Polak, 1957). It has two chief building blocks. The first is an accounting identity, in which the change in the stock of money is shown as the sum of changes in its international and domestic components:

$$\Delta M = \Delta R + \Delta D \qquad (1)$$

where $M$ is the stock of money, $R$ is the local-currency value of the net foreign assets of the banking system and $D$ is the net domestic assets of the banking system, or domestic credit.

The second building block is an assumption that there is equilibrium in the money market, so that any change in the demand for money, $\Delta Md$, is matched by an equal change in supply, $\Delta M$:

$$\Delta M = \Delta Md \tag{2}$$

The demand for money is taken to be a function of changes in real income, the behaviour of which is assumed to be uninfluenced by monetary variables, and in the domestic price level:

$$\Delta Md = f(\Delta Y, \Delta P) \tag{3}$$

If we take the change in international reserves (net external assets) as the key BoP indicator, these three equations can be combined in a fourth which shows that BoP deficits (reserve losses) will be the result of increases in domestic credit in excess of increases in demand for (= supply of) money:

$$\Delta R = \Delta M - \Delta D = f(\Delta Y, \Delta P) - \Delta D \tag{4}$$

Although monetarist, the model can be rewritten in more Keynesian terms, in which excesses of absorption over income which result in BoP current account deficits can similarly be related to excess domestic credit creation.[2] It can be modified in a variety of ways. One is to disaggregate $\Delta D$ into credit to the private and public (or government) sectors, which draws attention to the potentially large importance for the BoP of fiscal policies and government borrowings from the banking system.

Although the financial programming model, and the monetary theory of the BoP generally, has become a good deal more sophisticated than as presented above, equation (4) presents the core of the IMF approach, i.e. as seeing excess domestic credit creation (and often underlying this, excessive government deficit financing) as the chief source of BoP difficulty. On the other hand, it is not accurate to characterise the IMF approach as purely monetarist (see Bird, 1984: 87–8; *The Quest*, pp. 216–20 for further elaboration of this argument). For example, and by contrast with the more 'academic' monetarist models associated with Frenkel and Johnson (1976), it is interested in the *composition* of the BoP and in policies to influence the behaviour of the current and capital accounts, and the basic model above can be modified for this purpose. The Fund also regards the exchange rate as an important direct influence on BoP performance, whereas monetarists of a more purist bent regard this as irrelevant except as it affects the balance between the supply of and demand for money. More elaborate versions of the model can incorporate the effects of exchange-rate changes, with policy solutions derived through iteration (IMF, 1987: 41).

## How the model is used

When a country requests BoP assistance the Fund staff will take a view of its monetary condition, and the extent of any excess of $\Delta D$ over $\Delta M$, using equation (4) or some variant of it. However, this basic calculation will be modified by the results of various sub-models, relating to import demand, the likely behaviour of export quantities and prices, feedback connections between the budget and the exchange rate, estimates of the likely volume of aid and other capital receipts, and so forth. There will also be a good deal of to-ing and fro-ing about the accuracy of the data, which often become bargaining chips in negotiations between Fund staff and the government (Martin, 1991: Chapter 2). And there will be a good deal of judgemental 'adjustment' of model results in response to the representations of the government's negotiators, the mission's common sense – where this differs from the model results – and the degree of pressure upon it to agree a programme. As shown in Chapter 2, there will also be consideration of policy instruments which go well beyond the basic model and address supply-side weaknesses, particularly in Extended Fund Facility (EFF) and Enhanced Structural Adjustment Facility (ESAF) programmes.

In short, it is not the case that the basic financial programming model is usually applied in a simple, mechanistic way. However, the lack of transparency in the lending policies of the Fund makes it impossible to assess the extent of its flexibility in practice. Equation (4) does still represent the analytical core of Fund programmes. Everything else is modification. This is most evidently the case with stand-by programmes, the content of which, as shown in Chapter 2, has changed only modestly over the years.

Essentially the same calculations, and the resulting domestic credit performance criteria, or 'benchmarks', remain at the core of even the less short-term and more supply-side EFF and ESAF programmes. The more 'structural' policies of these have been grafted on to traditional performance criteria, and have not substituted for them.

It is therefore important to examine the qualities and limitations of the core model when forming a view of the appropriateness of the Fund approach to programme design.

## ISSUES ARISING

### Underlying assumptions and attention biases

To ask about the realism of its underlying assumptions is a good first step in the evaluation of any formal model. The financial programming model is open to a number of criticisms from this point of view. One source of weakness, as the Fund admits (IMF, 1987: 22), is the assumption that the demand for money is stable and predictable. This condition is often not satisfied, with a general tendency for the income velocity of circulation ($v$) to vary inversely with $\Delta M$. Of course, if the variability of $v$ in response to $\Delta M$ is sufficiently well understood and consistent, allowance for the forecast change in $v$ can be incorporated in the calculations of 'allowable' $\Delta D$. This is sometimes done. However, such adjustments introduce a further element of judgement in the mission's calculations, reducing the 'objectivity' of its results; in most programmes the standard assumption of a constant $v$ is retained, because of the difficulties of forecasting its behaviour during the programme period.

This brings us to the assumption that the demand for money, $Md$, is independent of $\Delta D$, i.e. that $\Delta D$ does not affect either real incomes or the price level. As the Fund staff have again pointed out, such an assumption will be valid only in restrictive circumstances. In most real-world situations we must expect changes in the level of domestic credit to influence the level of economic activity, for example by influencing the availability and cost of working capital, and the demand for, and prices of, non-traded goods and services.[3] Here too Fund missions sometimes attempt to make allowance for such interconnections when determining their credit ceilings, particularly in 'growth-oriented' programmes, but the extent of any such adjustments is in practice commonly relegated to the central objective of reducing the BoP deficit (G-24, 1987: 13). In any case, the introduction of such adjustments further reduces the 'objectivity' of the numerical results.

As will be shown later, it is also open to doubt whether many developing-country governments are actually in a position to exercise the degree of policy control over $\Delta D$ that the financial programming model requires. Moreover, the model can be criticised for the exclusive way in which it focuses on aggregate demand and domestic credit as *the* source of BoP difficulties. Often, of course, they result from adverse terms of trade or other external developments, as shown in Chapter 3.[4]

Even among domestic sources of difficulty, while excess demand often *is* a cause of payments deficits, structural weaknesses on the supply side of the economy are a no less common cause in developing countries – a fact acknowledged first by the World Bank when it opened its 'structural adjustment' lending window at the beginning of the 1980s and subsequently by the Fund with the introduction of its SAF and ESAF credits. For example, the financial programming model takes exports to be exogenously determined by world economic conditions and other factors external to the model – a paradoxical position, given that Chapter 3 found that the chief positive impact of Fund programmes is on export volumes. In this area, then, the model gets in the way of, or at least makes little contribution to, the preparation of realistic corrective policies in the common situation of supply-side weaknesses. To be fair, the Fund almost invariably includes devaluations in its programmes and sees the exchange rate as a policy instrument essentially directed to altering the structure of the economy, as between tradables and non-tradables (IMF, 1987: 36–9). But, so perceived, this is not a policy instrument which can readily be incorporated into the basic financial programming model.

Even in situations where excess demand is the chief source of difficulty, the model is open to the criticism that, by focusing on quantified aggregates, it diverts attention from qualitative aspects of policy. This is an implicit line of criticism developed in the Fund's own Fiscal Affairs Department and we shall return to it shortly. The Fund is now paying more attention to the 'quality' of fiscal adjustments; but the financial programming model does nothing to help it, and it took at least a quarter-century for this limitation to be reflected in Fund orthodoxy.

The financial programming model is also unsatisfactory in the way in which it handles time. Its critics describe it as static, but the Fund's staff (IMF, 1987: 20–1) say it works best in the long run. Both are right but both draw attention to further defects. It is a 'long-run' model in the sense that it requires sufficient time to elapse for all adjustments to work themselves out, in order to validate the assumption of equilibrium in the money market underlying equation (2). In the short term there can be no presumption of equilibrium. How long is long-run? The Fund staff think that the adjustment lags are likely to work themselves out over the period of a stand-by programme, typically 12 to 18 months, while acknowledging that it will sometimes take longer.[5] Provided most adjustment lags work themselves out within this type of period, the time factor in the financial programming model appears to provide a basis for calculating credit ceilings – except that it is used for calculating quarterly ceilings, often beginning three months after the start of the programme. The model provides no assured basis for such calculations.

While the model is 'long-run' in the sense just described, it is not dynamic and has been criticised for its inability to cope with important time lags, with uncertainty and the formation of expectations.[6] Edwards (1989b: 19–21) has particularly criticised the model along these lines, suggesting that, in consequence, the Fund has not kept abreast of best practice in the design of macroeconomic policies, e.g. in incorporating the effects of private sector reactions to government monetary policies, and in maximising the credibility of these policies:

> [The financial programming] model has failed to formally incorporate issues related to the inter-temporal nature of the current account, the role of risk and self-insurance in portfolio choices, the role of time consistency and precommitments in economic policy, the economics of contracts and reputation, the economics of equilibrium real exchange rates . . . and the theory of speculative attacks and devaluation crises, just to mention a few of the more important recent developments in international macroeconomics (p. 20).

While acknowledging that such criticisms may sound rather carping or academic, he goes on to illustrate ways in which the new developments could be utilised to strengthen the basic financial programming model.

This static nature of the model has caused the Fund particular difficulties since it was pushed in the later 1980s towards the adoption of 'growth-oriented' programmes. As its staff have pointed out (IMF, 1987: 27–8), the incorporation of economic growth as a policy objective alongside the BoP generates a host of complications, bringing in several new variables and relationships, with complex and lagged interactions. This reduces confidence in the underlying parameter values and increases the difficulties of implementing the model for the purposes of policy formation. As a pragmatic solution, where the growth objective is given weight it is accommodated by more or less *ad hoc* adjustments to the BoP target value, to allow a larger volume of imports – a procedure that again undermines the 'objectivity' of the resulting performance criteria.

The financial programming model, like any other economic model, takes a particular view of the characteristics of an economy. Some of its behavioural assumptions have been mentioned already. The neo-structuralist school, most closely identified with the name of Lance Taylor, identifies a considerably larger number of variables whose behaviour is viewed as often differing from the postulates of the Fund. Taylor (1988: 148–54) suggests that:

(a) Many prices are determined by a fixed mark-up rule rather than by the 'flexprice' market-clearing assumption underlying the Fund view. Where mark-up pricing is prevalent, inflation is more likely to emanate from supply-side weaknesses than from excess demand.
(b) Stabilisation-related credit restrictions and demand contraction are liable to cause reductions in output, employment and real incomes (particularly induced by the reduced availability and/or higher cost of working capital), and not merely the adjustment of relative prices envisaged in monetary theory.
(c) The devaluation which is an ingredient of most Fund programmes is likely to have stagflationary effects, driving up prices by raising production costs while simultaneously absorbing demand through higher import and other prices. Although there may also be expansionary effects through increased output of tradables, there will often be a net stagflationary impact, contrary to the Fund model. In fact, the inflationary process it sets off, and the prevalence of supply-side bottlenecks which cannot be resolved simply by altering relative prices, diminishes the probability that the devaluation will stimulate the production of exports and import substitutes.
(d) Whereas the Fund sees public investment as crowding out private investment, so that reductions in the former may stimulate the latter, the neo-structuralists emphasise the

'crowding-in' effects of public investment, with improved infrastructure, communications and economic services stimulating private investment.

(e) Given the high import content of capital formation, there is little scope for substitution with locally produced capital goods, so that reductions in imports resulting from Fund programmes are liable to limit new investment in export and import-substituting activities, with the result that beyond the short term the net BoP effects may be small or negative.[7]

(f) The reductions in domestic absorption which result from Fund programmes may hamper export performance (rather than releasing resources for it) by reducing the ability of the domestic market to act as the necessary launching pad for a successful export strategy.

(g) The stagflationary effects of interest-rate increases (also commonly incorporated in programmes), absorbing demand and raising production (working capital) costs, are liable to swamp any efficiency-raising effects.

It will be suggested later that neo-structuralists appear to overstate considerably the stagflationary consequences of Fund programmes, but it is not necessary to believe that each of the above points carries great weight to conclude that they add up to a challenge to the orthodox view of economy-wide behaviour that cannot merely be dismissed. They point up a range of variables whose behaviour can vary from one economy to the next – and the dangers, therefore, of taking any uniform view of macroeconomic processes.

It is evident from the above that the Fund model has significant limitations and is open to a range of criticisms. Before considering the implications of these, however, we shall briefly examine some outstanding issues surrounding the chief policy instruments employed in its programmes.

## Doubts about the policy instruments

### *The exchange rate*

As was shown in Chapter 2, devaluations have become an almost invariable component of Fund programmes, except in currency-union countries where the exchange rate is not available as a policy instrument. In the 1960s and 1970s devaluation (although then a less common programme ingredient) was among the most fiercely debated of the Fund's policy stipulations. A good deal of the heat has gone out of this controversy as more countries have moved from fixed to flexible exchange-rate arrangements, but to the traditional arguments has more recently been added what might be labelled a 'conservative' critique, as described in the companion volume (Bird, 1995: Chapter 2). It is a critique which the Fund is obliged to take seriously because it is espoused by some of its most important shareholders, most notably Germany and some of the other European Union countries (for a brief espousal of this argument see Sachs, 1989a: 114–15).

The argument here, echoed in debates about the European Exchange Rate Mechanism, is that commitment to a fixed nominal exchange rate serves as an anti-inflationary anchor, avoiding the price-raising effects of currency depreciations and *obliging* the government to pursue the fiscal and monetary rectitude necessary if the fixed nominal rate is not to appreciate in real terms, to the detriment of the BoP. By thus committing the government, the fixed rate is seen as adding credibility to its anti-inflationary stance, reducing inflationary expectations. Conversely, exchange-rate flexibility can be seen as undermining fiscal discipline, permitting excess demand to be accommodated, and adding, both directly and indirectly through its effects on public expectations, to inflationary pressures. Full flexibility can be seen as a form of indexation, bringing with it all the dangers of a self-perpetuating inflationary cycle associated with other forms of indexation.

Here the conservative and neo-structuralist arguments tend to merge, for both (if for different reasons) doubt governments' ability to make nominal devaluations 'stick' in real terms and hence their ability to bring lasting benefit to the BoP. In the neo-structuralist case, to this critique is added elasticity pessimism, because of expected severe (non-price) supply-side production constraints and the claim already mentioned that devaluations tend to be stagflationary. Neither school denies that countries with seriously overvalued currencies need to attend to their exchange rate but, in the absence of large misalignments, both appear to hold that a fixed nominal rate is superior to a flexible one.

Against this, the Fund can mount a strong defence, however.[8] At the level of principle, it can point out that if we see the exchange rate as a 'switching' device, altering relative prices in favour of tradables, there are no satisfactory alternatives, for using taxation to achieve similar results is (a) difficult in practice and (b) induces additional distortions. It can also be objected that the view that a fixed exchange rate imposes fiscal and monetary discipline wishes away all the political and practical difficulties of achieving the macroeconomic control necessary to make a fixed rate workable. With powerful governments such as those of Britain and Spain unable to achieve such outcomes, what chance is there for many of the more fragile regimes of the Third World (and Eastern Europe), operating with weaker fiscal and monetary instruments through more rudimentary financial systems? The Fund can also point to the difficulties created for African Franc Zone countries by the fixity of their exchange rate with the French franc between 1948 and early 1994. As the French franc became a relatively 'hard' currency, effectively tied to the Deutschmark, this forced substantial deflationary costs on the Franc Zone states without avoiding overvaluation and its attendant problems, thus leading to the eventual collapse of the parity and a doubling in the CFA/French franc rate in early 1994.

At the empirical level, the Fund can point out that the evidence does not bear out the worst fears of either the conservative or the neo-structuralist critics. We saw in Chapter 3 not only that Fund programmes are associated with substantial real reductions in the exchange rate but that these reductions are sustained at least into the medium term. This result was consistent with the findings of Kamin's (1988: 28–9) survey of the empirical evidence, that there was no evidence that the additional inflation caused by devaluations is sufficient to cancel out real depreciations. Edwards (1989a: Table 1) similarly found that a substantial proportion of nominal devaluations in developing countries 'stuck' in real terms (although there was a significant minority of devaluing countries where this was not achieved). In a separate study (1989b: 45–7), he also casts doubt on the argument about the stagflationary effects of devaluation, finding that, although there are short-term contractionary effects, these are not sustained and, in any case, that the measures recommended as alternatives to exchange-rate action are also liable to have stagflationary consequences.[9] Kamin (1988: 29–31) found no evidence that devaluations have a direct contractionary effect. Unpublished IMF staff reviews find positive export responses to devaluations (especially by non-traditional exports),[10] and lack of correlation between exchange-rate flexibility and lax financial policies.[11] They and others can cite cases where sharp devaluations have been accompanied by *reduced* inflation rates (see, for example, Kimaro, 1988: 17; Quirk *et al.*, 1987: 29–30).

However, while the stronger criticisms of Fund use of devaluations appear not to be sustained in the general case, this is not to deny that there are circumstances when one or other of the criticisms will be validated, and that there is a degree of risk in most country situations that nominal devaluations will be undone by consequential price rises, which do little for the BoP but accelerate inflation. To this caution might be added the 'fallacy of composition' (or 'adding-up') argument. This suggests that deployment of devaluation in large numbers of developing countries comes near to Fund instigation of a process of competitive devaluations (quite contrary to its objectives), with particular risks for primary product-exporting countries that the aggregate effects of the resulting increases in export supplies will cause self-defeating reductions in world prices. While the evidence on this 'immiserisation' hypothesis is still tentative, it appears that it has been an important factor for exporters of cocoa and, perhaps, coffee but that there has been no large depressing effect on commodity prices in general (World Bank, 1994a: 45–9). Indeed, among African countries the problem has rather been the contrary one, of a small quantity response to improved price incentives (see Cleaver, 1985; Diakosavvas and Kirkpatrick, 1990; Faini and de Melo, 1990; Jaeger, 1991, all of whom report only limited responsiveness of African agriculture to devaluations).

All these considerations point to the desirability of pragmatism and selectivity in the employment of this instrument.

### *Monetary policy*

Although the analytical approach of the Fund is not narrowly monetarist and is fairly eclectic, none the less its programmes do centre around the deployment of monetary policy, chiefly directed at the control of domestic credit. But monetarism, and monetary policy, has gone through rough times in recent years, with a disillusionment in industrial countries arising from the large claims made for it in the 1970s.

The reasons need little rehearsal. How money should be defined and what measure is the most useful for policy purposes have proved elusive, not to say insoluble, problems. The monetarist expectation that credit restrictions would have few or no adverse consequences for output and employment has been decisively disproved, thus reducing the political attractiveness and sustainability of strict monetarism. The internationalisation of financial markets and the rapid pace of innovation in the financial sector (besides further reducing the stability of the demand-for-money function) have reduced the practicability of achieving the desired control over money supply within the boundaries of a national economy, and banks have proved resilient in resisting the desires of the monetary authorities when these seemed contrary to their own interests (see Stiglitz, 1992: 300). As these difficulties have made themselves felt and governments have turned away from exclusive reliance on monetary instruments, so the professional consensus – never strong[12] – has been eroded.

But if the practicability of the monetary approach is under question in the conditions of industrial countries, what of the developing (and indeed East European) countries to which Fund conditionality is actually applied? We can start with the following admission of the Fund staff (IMF, 1987: 9–10):

> the choice of policy instruments is heavily influenced by the stage of development of economic institutions. In a country with sophisticated financial markets, for example, there are more means available for the government to influence the rate of monetary expansion (although there are also more ways to satisfy the demand for credit, in the face of restrictive official policies, through the layering of financial assets). In a country with a relatively undeveloped, sharply segmented, financial market, the economy is likely to respond much less flexibly to changes in monetary policy. Moreover, where there are severe policy-related distortions – arising from price controls, exchange and trade restrictions, overvalued exchange rates, and official ceilings on interest rates – the efficacy of normal demand-management policies is greatly weakened, and the need for structural changes is all the more urgent.

This comes close to an admission that the Fund's chosen policy instruments are best suited for the 'developed' countries, which refuse to borrow from it, and least suited for the low-income countries of Africa and elsewhere, where the Fund has a high proportion of its programmes.

Doubts about the suitability of the Fund's stress on the control of money and credit are reinforced by the difficulties in developing country conditions of achieving the required control (see Healey and Page, 1993, for a useful summary of the uses and limitations of monetary policy in developing country conditions). The policy variable that can be used to 'solve' equation (4) is domestic credit, $\Delta D$: given the expected values of $\Delta Y$ and $\Delta P$ and a target value of $\Delta R$, it is assumed that the necessary value of $\Delta D$ can be ensured by the monetary authorities. But this is a strong assumption, even if we are willing to allow that the desirable value of $\Delta D$ has been accurately estimated. It requires that the central bank, through open market operations, variations in commercial bank reserve-ratio requirements, quantified credit maxima or other means, should be able to achieve a rather precise control over bank lending to the private sector without inducing a movement into near-substitutes. It also requires that the government has a sufficient flow of accurate, timely information about its revenue and expenditure trends and prospects (and those of the parastatal sector) that it can reliably estimate its deficit financing requirements, and has a sufficiently exact control of these to keep them to a level consistent with the required value of $\Delta D$. The Fund's approach further requires that it is possible to define $\Delta M$ and $\Delta D$ satisfactorily so as to include all variables which have the essential qualities of that elusive property, 'moneyness'.

In many developing countries all these requirements are liable to be breached; at best, the level of control is only approximate.[13] In short, $\Delta D$ may have fewer of the attributes of a policy instrument, to be varied at will to achieve a given payments target, than the financial programming approach presumes. These considerations may help to explain the weak revealed ability of programmes to limit the actual expansion of $\Delta D$. In developing country circumstances, when open market operations and other alternatives are unfeasible, the most reliable way of holding $\Delta D$ within a programme ceiling is through quantitative controls on commercial bank credit. Here too the Fund approach can be criticised as having undesirable biases:

(a) Quantified ceilings on commercial banks are liable to be inefficient, for the banks will give priority to meeting the ongoing needs of existing customers, and the larger customers among those, to the disadvantage of the financing of innovations and new businesses.

(b) Such ceilings reduce banks' incentives to liberalise, by reducing active competition for new customers.

(c) By focusing on quantified targets, the Fund's credit ceilings may also be inappropriate to financial liberalisation, which operates on the demand for and supply of money chiefly through interest-rate policy.

None of these complaints is aimed at the truism that the avoidance of large-scale excess money creation is essential to macroeconomic stability. The issue, rather, is whether the Fund's way of going about things is the best or whether it is not too narrow in the policies deployed in its programmes.

*Fiscal policies*

One line of defence would be to argue that the above criticisms miss the point that the central thrust of the Fund approach is actually directed to fiscal policy. Indeed, Sachs (1989b: 113) has argued that 'The Fund's emphasis on fiscal policy mismanagement as the key source of BoP problems is its main strength and is indeed the core "truth" of its strategy.' Others have suggested that the Fund pursues an essentially fiscal, rather than monetary, approach to the BoP.[14]

The Fund's past approach to fiscal policy is also open to major criticisms, however. The chief of these, remarkably, has been developed within its own Fiscal Affairs Department, under the influence of its Director, Vito Tanzi.[15] This agrees that fiscal policy is central to BoP management but disputes the usefulness of focusing on the size of the budget deficit, on the grounds that it diverts attention from the real problem – namely, the 'quality and durability' of the specific fiscal measures used by the government to remain within the desired budget ceiling. Excessive budget deficits are often accompanied by distortionary tax systems and inefficient expenditure patterns, so that merely reducing the deficit will not go far enough. To make matters worse, when faced with a necessity to reduce budget deficits, governments often seek to minimise political costs by inefficient means: cutting disproportionately on capital formation; starving civil servants of the supporting inputs they need if they are to work productively; executing across-the-board cuts with little heed to comparative economic costs and benefits. Moreover, presumably on the basis of the Fund's experiences, Tanzi suggests that 'the longer ceilings on macro variables are in use, the more ways countries learn to get around them' (1989: 21).

On this view, the connection between the BoP objective and the budget is sensitive to precisely how the government raises its revenue and trims its expenditures. Therefore (Tanzi, 1989: 24):

> provided that a country is willing to implement considerable structural measures early enough in a program so that the positive effects of these measures can be felt relatively soon, the Fund should be prepared to require less reduction in the overall deficit . . .

To some extent, Fund practices have taken this critique on board, for it was shown in Chapter 2 that it has been taking a considerably more detailed interest in the composition of fiscal policy than was formerly the case. Resistance to this trend is considerable, however, so that according to Tanzi (1989: 25):

> A perusal of stabilization programmes indicates that despite an increasing awareness of these issues, political difficulties, guidelines on conditionality, and timing concerns have prevented their being taken formally into account in Fund programmes.

This situation has reportedly improved since then, but one suspects that in the heat of programme negotiations and under pressure to reach agreement, the finer details of fiscal policy are still sometimes subordinated to preoccupation with the numbers which should make up the performance criteria.

If so, this is also regrettable for a related reason, which concerns the complexity of the connection between the fiscal balance and the BoP. The literature on 'fiscal stance' and the economic effects of budget deficits has shown the many interconnections between fiscal and other economic variables, complications which make it difficult to predict the magnitude of fiscal change necessary to achieve a given BoP (or other macroeconomic) target. Indeed, Buiter (1985: 54) asserts that

> there are no 'model-free' measures of fiscal impact on aggregate demand. Different views on how the economy works

> will give rise to conclusions about the demand effect of fiscal policy measures . . . that may differ not only in magnitude but even in direction.

Finally, it can be asked whether the Fund's programmes actually create the 'hard' budget constraints they are assumed to do and whether, in consequence, they are able to achieve the reallocation of credit in favour of the private sector which the Fund desires. Reviewing the literature on programme effects, one is struck by evidence that little such reallocation occurs. Thus, Ground (1984: 81) found that, 'contrary to all expectations', restrictions on credit within Fund programmes appeared more stringent for the private sector than for the public sector. Our own investigations of programme effects found no significant reduction in the share of total credit going to central government and some slight (non-significant) tendency in the other direction (Chapter 3, Table 3.5). A more in-depth study of the Kenyan case similarly found that Fund programmes made more generous provisions for credit to the government than for total domestic credit (Killick and Mwega, 1993: 59). Since the Fund would like to shift credit in favour of the private sector, it may be speculated that it is unable to achieve this in the face of fiscal and political realities. If so, it is among the best-kept secrets of a secretive organisation.

More hard evidence on credit shares would be necessary before arriving at firm conclusions. If indeed there is a systematic tendency for the government, or public sector, to be relatively favoured, this would suggest an inability on the part of the Fund to secure implementation of tough deficit-reducing measures.

## The indefensibility of performance criteria

If we take the combined weight of the earlier critique of the basic financial programming model and the above comments on the chief policy instruments employed, the conclusion seems inescapable *that quantified credit ceilings and other performance criteria are literally indefensible.* Recall first how performance criteria work. The values for the external asset holdings of the central bank, for credit ceilings, for reduction in fiscal deficits and the like are initially generated by application of the financial programming model. These values are then modified in the light of the Fund mission's judgements and negotiations with the government. Sometimes the underlying BoP projections are manipulated to result in 'acceptable' (but unrealistic) figures.[16] These are then quantified on a quarterly basis for stand-by programmes (six-monthly in the case of ESAFs), with the government's continued access to the credit conditional on remaining within them. If they go above the ceilings the programme is either discontinued or (usually after a delay) the Fund agrees a waiver of the original conditions and access is resumed.

But what claim can a quarterly ceiling on credit to the government have to be objectively derived from economic analysis? And how much confidence can be felt for the 'correctness' of the resulting number? Clearly it is not objective, not only because it is an outcome that has to be negotiated (perhaps initially within the Fund team – with country specialists tending to desire different outcomes from the representative of the Policy Development and Review Department – and then not only between the Fund and government teams but also with the World Bank) but also because it is based on a myriad of judgements – about the behaviour of $v$; the response of the BoP, and hence $\Delta R$; the consequences of the currency devaluation which is also likely to be included in the programme; the behaviour of GDP and prices, determining $\Delta M$; feed-backs between the budget, the BoP and other variables and the resulting desirable value of the fiscal balance; the many time lags involved in these processes; and so on. The resulting figure can inspire little confidence, for although it is desirable that Fund staff should be allowed to exercise their individual judgements in putting the programmes together, the cumulative effect of such a string of judgements (or negotiating compromises) can be large.

To these considerations we can add: unreliable data; the possibilities described earlier that the economy will behave in different ways from those postulated by the Fund; the difficulties of tracing the links between the fiscal balance and the BoP; the variable effects of detailed tax and spending decisions; the use of a 'long-term' model to derive short-term ceilings; and the conceptual and practical difficulties of keeping $\Delta D$ within the desired range. The numbers written into programmes, *determining access to the Fund's BoP assistance*, are just not to be taken seriously.[17]

A possible rejoinder by the Fund would be that it is flexible even in its use of quantified performance criteria. It can point out that nowadays it makes greater use of review missions as an alternative to predetermined performance criteria, so that it is easier to adjust ceilings that appear to have become less appropriate in the light of changing conditions, and it can leave the determination of some performance criteria until a later stage, thus reducing the potential severity of forecasting errors.

Second, it can point out that it uses waivers (see *The Quest*, p. 202) – releasing the government from the ceilings written into the original programme – so as to provide greater flexibility and avoid the problems created by ceilings that turn out to be unfeasible. Indeed, the limited evidence available indicates that the Fund has frequent recourse to waivers. Martin (1991, Table 2.6) records that of 95 programmes begun in sub-Saharan African countries in the period 1980–86 no less than 78 were subsequently the subject of waivers. He goes on (1991: 284–5) to report an even higher incidence of waivers in more recent programmes. The necessity for these arose chiefly out of difficulties with performance criteria relating to budget deficits and domestic credit – precisely the areas where our earlier arguments predict problems in the setting of meaningful targets.

While the greater use of review missions is to be welcomed, the availability of waivers is a weak defence of quantified performance criteria for a number of reasons. If it is necessary to grant waivers other than exceptionally, this is evidence that the quantification process is unsatisfactory. Otherwise, why should waivers be needed so frequently? The fact that waivers are apparently necessary in a large majority of African programmes indicates that the Fund's standard approach does not work well, at least in this large region.

Unfortunately, the Fund's policies and practices on waivers are shrouded in mystery. It is unsatisfactory that this important aspect of the Fund's operations should so lack transparency, leaving governments uncertain about where they stand and about the rules to which they are supposed to conform. These uncertainties are all the greater because readiness to grant waivers has been used by the Fund in the past as a policy instrument, a tap to be turned off or opened wide according to the *global* circumstances of the time.[18] The Fund denies that it still uses waivers in this way but reportedly it has recently (1994) made the granting of waivers easier as a way of keeping programmes on track in the face of unforeseen shocks – a *policy decision* which could just as easily be reversed again.

The situation is made more unsatisfactory by the Fund's practice of suspending access to a credit pending a decision on whether to grant a waiver and agreement on new performance criteria. This withdrawal of access, even if only temporary, can cause severe disruption to governments in often desperately tight BoP situations, undermining the credibility of the programme and the likelihood that it will stimulate capital inflows and investment.[19]

In short, as currently practised waivers are an unsatisfactory response to the difficulties created by quantified performance criteria. The position would be improved if the Fund were to introduce greater transparency into its policies on waivers, but to do so would lay bare the extent of its difficulties with quantified criteria. If these are to be justified at all, it can only be on the pragmatic grounds that it is impossible to think of any better way of proceeding. But before taking up that challenge there are further issues to be examined.

## Dealing with growth

That Fund programmes tend to depress economic growth and impose avoidable economic hardships are long-standing complaints. We have already reported some of the criticisms of Taylor (1988), who emphasises the stagflationary potential of the programmes and the Fund's past lack of interest in their distributional consequences. Ground (1984: 80) is another critic in the neo-structuralist tradition who emphasises the deflationary and poverty-increasing nature of many of the Fund's policy preferences, arguing that Fund programmes contain five types of 'recessionary bias':

> [1] the insufficiency of the amount of financing; [2] the inconsistency of domestic economic policies; [3] the use of the stock of net external assets as a performance criterion; [4] the use of specific fixed targets for the performance criteria; and [5] the nature of the link between external financing and adjustment agreements.

One specific suggestion (p. 81) is that the difficulties which borrowing governments experience in remaining in conformity with performance criteria may cause them to play safe by adopting policies that are more contractionary than is necessary. The late Sidney Dell (1982) complained of a 'political economy of overkill' and the Group of Twenty-Four (1987: 19–20) similarly talk of a 'built-in tendency for domestic credit to be tightened excessively as a result of unrealistically low projections of the inflation rate. The consequential overdose of credit controls leads to output contraction . . .'

Against these complaints we should set the conclusions of the literature surveyed and the tests reported in Chapter 3 on the effects of Fund programmes. Most of the tests reviewed there indicate that Fund programmes are not associated with any significant loss of output; our survey also failed to uncover evidence of large distributional, political or supply-side programme costs. Nor did we find much association between programmes and inflation. In short, although the evidence is not conclusive, it is easy to exaggerate the likely extent of adverse stagflationary and poverty consequences. However, the evidence presented did unambiguously link IMF programmes with substantial and sustained declines in investment rates. In the short to medium term, such declines may not translate into output losses because of compensating improvements in the efficiency of resource use, but it cannot be expected that moving the economy closer to its efficiency frontier could continue to compensate for reduced investment over an extended period.

The jury thus remains out on whether programmes have growth-reducing consequences. Even if they do not, it is possible that a different approach would produce better results. We should recall here the work of Khan and Knight (1985: Chart 1) showing substantially faster growth for countries adopting programmes which include supply-side measures, as against a conventional stand-by demand-management approach – a result consistent with internal Fund reviews which indicate relatively positive growth results from the more 'structural' SAF and ESAF programmes.

In an earlier paper, Khan and Knight (1982) point out that different combinations of stabilisation and structural adjustment measures have different effects on growth and other variables, and the importance, therefore, of searching for cost-minimising combinations. This is a theme also taken up in *The Quest* (Chapter 8) which argues the case for the adoption of a cost-minimising approach to programme design and criticises the Fund for having neglected this in the past.

Unfortunately, the situation in this regard has changed little during the past decade. While Chapter 2 has shown that the Fund does now give the growth objective slightly greater weight in some of its programmes, and that its SAF and ESAF facilities have been important innovations, the discussion of ESAF programmes in Chapter 3 (pp. 76–85) suggests that there is much pretence in Fund claims to give higher priority to growth in these programmes. Its missions still do not go about programme design within a cost-minimisation framework. The achievement of some minimum growth rate is still not accepted as a constraint on programme design. In contrast with the approach of the World Bank, Fund missions base their BoP projections on estimates of the likely availability of external resources, with growth a residual outcome, rather than estimating the volume of support needed to achieve a desired level of economic activity.[20]

Like its financial programming model, then, the Fund's approach remains essentially static or short term. Would it be possible to modify the model to make it more dynamic, incorporating growth as a target variable? There is reportedly little interest in such a project among the Fund's country staff, although there is no reason in principle why this should not be done. Khan, Montiel and Haque (1990), for example, have produced a formal integration of the financial programming model and the modified two-gap ('RMSM') model employed by the World Bank which endogenises GDP growth along with the BoP and inflation as target variables. However, they point out that the resulting model can be kept reasonably simple only through resort to some rather drastic assumptions, and that making it more realistic would quickly increase its complexity, thus reducing its operational value. The Fund itself (1987: 27–8) has stressed the complexities involved in building a growth objective into the financial programming model, describing it as 'a formidable task' which researchers have only just begun to undertake; and Mohsin Khan, one of the principal researchers involved, has consistently stressed our limited knowledge of the connections between the key components, e.g. between financial variables and the real economy.

The solution favoured by the G-24 is for Fund missions to undertake 'growth exercises' (1987: 16):

> In order to provide Fund programmes with a growth perspective, it is proposed that a set of 'growth exercises' be performed prior to the 'financial exercises.' From these exercises, the amount of external finance necessary to support a growth-oriented adjustment programme could be

> determined. The financial exercises should be built upon these estimates of necessary external finance.

Taylor (1988: 163) favours a similar approach.

We shall consider a fundamental snag with such proposals in a moment. But before doing so, we can draw the conclusion from the last few pages that the Fund's traditional financial programming model copes poorly with the behaviour of the real economy over time, and makes no contribution to the reorientation of the Fund towards 'adjustment with growth'. Its staff play down the significance of this deficiency, arguing that in practice programme design is based on a far wider range of considerations and that programmes can be adjusted in *ad hoc* ways to accommodate growth. However, the necessity for such *ad hoc*ery adds to the list of respects in which the financial programming model appears seriously flawed.

When considering the modification of the Fund's traditional approach to a cost-minimising, growth-promoting one, we should recall the short-term nature of its programmes. It is a further long-standing complaint that these are too brief to be able to address growth and supply-side weaknesses. Although earlier chapters have recorded the extent to which the Fund has moved towards medium-term programmes and that it often permits a government to enter into a succession of programmes, these are no more than partial solutions. According to Louis Goreux, formerly Deputy Director of the Fund's Africa Department, the main reason for the frequent failure of its programmes in Africa was that supply responses were slower and weaker than expected. This was more important than policy slippages or external shocks. The Fund had thus found itself entering into successions of programmes and into 'structural adjustment' programmes in collaboration with the World Bank. However, as Goreux (1989: 146) says,

> Plugging a number of supply measures advocated by the Bank into the Fund model is not the appropriate solution. . . . Supply and demand considerations need to be integrated into the formulation of the model to determine the speed of adjustment that is both technically feasible and politically sustainable. [This speed] has often been over-estimated, partly because the length of the grace period attached to Fund purchases was too short.

In other words, missions find it necessary to build adjustment speeds into programmes which they know are unrealistic.

The standard IMF response to criticisms along these lines is to point out that its Articles of Agreement only permit it to make its resources 'temporarily' available to members with payments difficulties. Under the Bretton Woods settlement it was the Bank, not the Fund, which was envisaged as the agency for long-term lending. In the 1960s and 1970s 'temporary' was interpreted to justify the 12–18 months typical of stand-by programmes. However, this legalism cannot be the chief obstacle to programme lengthening, for 'temporary' is as long as a piece of string – and has already been stretched to accommodate three-year EFF, SAF and ESAF programmes. The Fund's willingness to agree long, virtually unbroken, successions of programmes similarly reveals the elasticity of this concept. There is no linguistic or legalistic necessity to take 'temporary' as synonymous with short term. It could equally persuasively be interpreted to mean *non-permanent*, in which case there would be no evident difficulty with programmes lasting five years, or even longer.

Money is the real obstacle, as Graham Bird (1995, Chapter 2) shows in the companion volume. Longer programmes and more extended transition periods require larger amounts of support, to finance the longer period before BoP viability is achieved. The G-24 (1987: 20) recognised this explicitly in connection with its suggestion of 'growth exercises':

> This implies a need for an effective acceptance by creditor countries of the concept of symmetry in adjustment and an obligation of those with large surpluses to provide the capital required – an obligation that would constitute, in effect, a reciprocal performance criterion for these countries.

Sadly, there has been a retreat from the norms on international financial co-operation over the last decade and a half. Surplus countries show no sign of accepting the obligations of which the G-24 reminded them. This unwillingness to put up more finance is the hole in the heart of all proposals to give the Fund a greater growth orientation, for while representatives of the industrial countries are happy enough to urge the desirability of 'adjustment with growth' they have a proven unwillingness to face up to its financial implications. At the same time, however, there is much wishful thinking in this unwillingness. For a succession of short-to-medium-term programmes is likely to add up to just as much financing as one or two longer-term ones, while being an intrinsically sub-optimal way of dealing with long-term problems. The creditor countries do themselves no favours by providing support in inefficient ways, not least because this contributed to the emergence in the 1980s of a serious problem of arrears to the IMF (see Chapter 2 and Bird, 1995, Chapter 1).

Nevertheless, industrial-country reluctance to enter into additional financial commitments is something with which the Fund has to live, as witness the difficulties it had with the 50 per cent increase in quotas agreed in June 1990 but only ratified at the end of 1992 (see IMF, *Annual Report, 1992*: 70–1). Even after ratification, the 'access' rules governing credit size were changed so that maximum credits were little larger than under the old quotas. In consequence of such constraints, staff missions find themselves having to write unrealistically short programmes, as Goreux reports, or manipulating payments forecasts so as to pretend that problems are not as large as they appear (see Martin, 1991: Chapter 2).

## The problem of 'ownership'

In evaluating its experiences with structural adjustment programmes, the World Bank has long suspected that the implementation of programme stipulations, and hence their likely impact, is strongly influenced by the extent to which the borrowing government regards the programme as its own – what the Bank has called the government's 'ownership' of the programme. A report by its Evaluations Department has taken this thinking a good deal further.[21] Assessing ownership by the extent to which the initiative for the programme's policies was local or external, the level of intellectual conviction in the appropriateness of its measures, the extent of support from the top political leadership, and efforts towards consensus-building among the wider public, it tested for correlation between this variable and its assessment of the satisfactoriness of programme outcomes. The results are reproduced in Table 4.1.

The results are strong and confirmed by various significance tests. Ownership was high in most programmes achieving good results and low in ineffective programmes. Of course, correlation should not be confused with causality and the Bank did not undertake causality tests. On the other hand, given the way the

*Table 4.1* Correlating programme outcome with borrower ownership

| *Borrower ownership* | *Programme outcome* | | | | |
|---|---|---|---|---|---|
| | *Highly satisfactory* | *Satisfactory* | *Unsatisfactory* | *Very unsatisfactory* | *Total* |
| Very high | 9 | 6 | 0 | 1 | 16 |
| High | 6 | 15 | 2 | 2 | 25 |
| Low | 4 | 10 | 6 | 3 | 23 |
| Very low | 0 | 3 | 7 | 7 | 17 |
| Total | 19 | 34 | 15 | 13 | 81 |

*Source*: Johnson and Wasty (1993: Table 1).

tests were structured, it is not clear how causality could have run from effectiveness to ownership, and the results are consistent with the findings of others who have studied the political determinants of programme success.[22] The Bank certainly interpreted the results to indicate that effective programmes are a *consequence* of borrower ownership, which was found strongly predictive of programme success in three-quarters (73 per cent) of all cases, with the most 'deviant' cases apparently explained by the intervention of exogenous shocks. The support of, or lack of opposition from, key interest groups was identified in the evaluation as probably the single most important influence. As the report states (p. 173), 'One of the most important services the Bank can provide is to ensure that the process of policy reform is "internalised" in the country as quickly as possible, so that the reform program is designed by the country itself . . .'.

Sadly, there is no equivalent information specific to the Fund. However, there must be a strong presumption that similar considerations apply to its programmes, not least because many of the Bank programmes analysed were accompanied by parallel Fund programmes. Its own tendency to attribute non-implementation to 'lack of political will' points in the same direction. So does the following summary of research into the influence of the IMF on the policies of the former communist countries of Eastern Europe (Henderson, 1992: 245):

> The IMF influenced the adjustment processes of several Eastern European nations in the 1980s through its efforts to promote market-oriented stabilisation and reform. Each country's state structure, policy-making process, and state/society relationships shaped its responses to IMF demands. Romania's

> centralist political system generated the most intense and successful resistance to the IMF's policies. Yugoslavia's polycentric political system, by contrast, weakened elite capacity to resist IMF demands while simultaneously impeding the implementation of IMF policies at the subnational level. Finally, Hungary's political system allowed the IMF to ally with elite supporters to promote its policies, yet also provided opportunities for lower-level actors to obstruct their implementation. Prospects for collaboration with the IMF have been enhanced by recent changes in Eastern European state structures and policy-making processes, which have encouraged freer political debate and more market-oriented development strategies. Yet impediments to collaborations remain, as political decentralisation has heightened the capacity of lower-level actors to obstruct standby implementation.

That the Fund has been unforthcoming on this subject is not, we suspect, because it thinks ownership is unimportant, but because it has particular difficulties in dealing with this subject.

Many of these arise from the crisis conditions in which governments often turn to the Fund, the intense pressures of work under which its country staff commonly operate, the speed with which its programmes are prepared and their often short-term nature. In such circumstances, with negotiating missions commonly lasting only two or three weeks, its staff do not have time to ensure that the government is fully 'on board', just as the government often will not have time (even when it has the desire) to undertake the consultations and public education necessary for consensus-building.

These intrinsic difficulties are not uncommonly aggravated by a certain arrogance of approach. Although we recorded in Chapter 2 some increase in IMF negotiating flexibility, including occasional willingness to settle for technically sub-optimal but politically more sustainable programmes, we also made it clear that this change was only marginal. The introduction of procedures for tripartite 'Policy Framework Papers' described there has been a useful device in the case of SAF/ESAF programmes. While it is widely conceded that in the early years the preparation of these was dominated by the Bank and Fund, there have been greater efforts subsequently to bring governments into the drafting process. That only limited progress has been made is, however, indicated by references in the IMF's 1991 *Annual Report* (p. 57) to the desirability of bringing governments more into the PFP drafting process.[23]

In any case, the key document for the Fund is the 'Letter of Intent' in which the borrowing government formally presents the policies it will undertake to strengthen the BoP and to promote other programme objectives. Herein, it might be said, lies the 'ownership' of Fund programmes. But these Letters, although ostensibly from the government, are still almost invariably drafted in Washington, with the government left trying to negotiate variations in a draft presented to it. It is difficult to imagine a procedure less likely to leave the government regarding the programme as its own. The practices of the Fund flatly contradict the obvious good sense of the World Bank report (1992: 15) that, 'One good indicator of ownership is the borrower's willingness and capacity to prepare the Letter of Development Policy' (the Bank's equivalent of the Letter of Intent). They do not encourage the desirable internalisation of policy reform mentioned earlier and, while it could be retorted that these practices ease the way for governments to use the IMF as a scapegoat, blaming it for unpopular measures they privately know to be inescapable, the Fund has become unhappy about being cast in such a role, on the grounds that governments ought to accept responsibility for the management of their countries' economies.

While acknowledging that Fund–government relations are nowadays rarely adversarial[24] and that the Fund is more sensitive than formerly to the need to bring governments into programme preparation, there is evidently a long way to go on this. It would be most valuable to have evidence for the IMF comparable to that presented in Table 4.1 on the Bank. Pending that, there remains a strong suspicion that a weak sense of ownership remains a major problem for the implementation of programme stipulations, and that this helps to explain the frequent programme breakdowns reported in Chapter 3.

So much for the critique. The next task is more difficult: to suggest a superior alternative. We turn to this in Chapter 5.

# 5

# CONCLUSION

## Is fifty years enough?

### THE CRITIQUES OF LEFT AND RIGHT

At the time of the fiftieth anniversaries of the IMF and World Bank in 1994, a group of voluntary organisations (NGOs) formed a campaign organisation called 'Fifty Years is Enough', attracting enough of the headlines to put a damper on the celebrations. Their campaign ranged more broadly than the concerns of this book but it is a convenient starting point for this concluding chapter to take up some of their complaints about IMF programmes.

The essence of these was that Fund (together with Bank) programmes were causing deflation in borrowing countries, widening income inequalities and worsening the welfare of the poor, jeopardising prospects for sustainable recovery and poverty reduction. Thus, according to spokesmen for the British charity OXFAM, the seemingly inevitable slide into dependency and despair can be reversed only when the defective and discredited policies of the IMF's adjustment policies in Africa are jettisoned for policies that do work; the Fund should be fundamentally reformed or extricated from Africa.[1] For some, at least, programme effects were so severe that the Bank and the Fund should be wound up (hence '50 years is enough'); others preferred the route of radical reform, including, for example, the phasing out of the ESAF. Underlying the discontent of these groups was opposition to, or at least deep scepticism about, the neo-liberal shift towards an increasingly globalised market-based world economy, backed up by the 'market-friendly' policy stances of the 'Washington consensus' – which is why it is appropriate to identify them with the political Left.

Also stimulated by the Fund's half-century, critics on the political Right were scarcely less sweeping. In their view, the Fund lost its *raison d'être* when the original Bretton Woods 'adjustable-peg' system broke down in the early 1970s. In its search for a role, the Fund gradually became yet another aid agency, channelling assistance from developed to developing countries (with the addition of Eastern Europe and the former Soviet Union at the beginning of the 1990s), losing its monetary character and straying into the design of medium-term 'structural' adjustment programmes in which it had little expertise. As Milton Friedman put it, 'With the collapse of Bretton Woods in 1971, the original function for which the IMF was established simply disappeared. But instead of closing down, the IMF turned itself into a junior World Bank.' For some of these critics too, abolition was the answer: 'the ideal solution would be to abolish the Fund and the Bank' (Walters, 1994: 22). For others the solution was for the Fund to return to its traditional short-term, monetary-based programmes, or to act largely as the provider of a seal of policy approval for governments which could then secure their financing from the private markets.

## An assessment

Those who have read Chapter 3 will realise that it takes a radically different view of the effects of Fund programmes from that of the critics of the Left. It found little evidence that Fund programmes are associated with any significant loss of output. It also showed that the distributional and poverty effects of Fund programmes were too complex, and too varied across countries, to permit sweeping generalisations. Among the poor there are gainers (often in the rural economy) and losers (often among the urban working class). Much depends on the pre-existing ownership of assets, economic and social structures, and the detailed design of adjustment measures.

Ironically, its critics flatter the Fund by imputing to it such dramatic effects. The truth is that, with exceptions discussed in Chapter 3, its programmes have rather limited revealed effects on developing country economies. How do its critics come to form such an exaggerated view? Leaving aside wilful distortion – by no means always absent – there are a number of reasons why this might come about.

First, that its programmes do not generally have large economic effects is not a defence that the Fund itself is anxious to promulgate. At least for the purposes of public discussion, it shares with its critics the premise that its programmes have large potency. It suits both sides to argue to this effect. Our findings in Chapter 3 suggest otherwise, however, with Fund conditionality described as something of a toothless tiger.

Another, very common, source of exaggeration arises from the use of dramatic anecdotal country-specific evidence. The survey in Chapter 3 (pp. 86–119) shows the wide variety of country experiences with Fund programmes, including some whose experiences were clearly adverse. A dramatic, but seriously misleading, story can be told by putting together a few of these – and by ignoring programmes that brought clearly beneficial results, or the majority where the outcomes were merely so-so.

The illegitimate attribution to Fund programmes of hardships which could more appropriately be blamed on the initial economic crisis (and on government policies which contributed to this) is another source of overstatement. The widespread belief that programmes result in particularly deep cuts in social spending is a case in point. As shown in Chapter 3, research on the incidence of government spending cuts finds that social services are actually among the more protected categories. While educational and health services have deteriorated seriously in Africa and Latin America in response to reduced government spending, Chapter 3 reports that most studies find no connection between these declines and the adoption of Fund or Bank adjustment programmes. More generally, the critics are apt to discount the effects of the initial crisis. Yet it is naïve to think that an economy which starts with some combination of inadequate import capacity, next to no reserves, an exhausted creditworthiness, a large monetised budget deficit, rapid credit expansion and serious supply-side bottlenecks can restore macroeconomic balance without hardship and unwanted expenditure cuts. And while in an ideal world the poor would be protected from these necessary reductions in absorption, the practical difficulty of achieving this protection, at least in the short run, is large, to say nothing of the indifference of a good many borrowing governments. In short, the critics tend not to come to terms with the problem of the counterfactual.

Relatedly, the critics have a tendency to argue in terms that tacitly deny, or at least discount, the importance of macroeconomic management, and to present adjustment as in opposition to development and the reduction of poverty. Again, the view taken here is quite different: namely that, properly understood, adjustment is a necessary condition for an effective long-run attack on poverty. In its broadest terms, adjustment should be seen as a response to a *permanent* need to keep abreast of changing patterns of world demand, output and competitiveness, technological change and unforeseeable shocks. No nation can sustain economic growth unless it can adapt, and in poor countries growth is indispensable if poverty is to be eradicated.

Macroeconomic management – the policy specialism of the IMF – is an essential part of this adaptation, responding to shocks and, above all, maintaining the stable economic environment which is so important for adequate responses to price and policy signals for change, and for the saving and investment necessary if an economy is to restructure itself in line with emerging economic and technological conditions. By contrast with the situation 25 years ago, the value for economic performance of avoiding large macro imbalances is no longer much contested within economics. There is ample evidence of the decisive influence of the policy environment as a determinant of a country's economic performance. The importance of tackling the foreign-exchange constraint, with all its implications for investment, capacity utilisation and human welfare, is self-evident. Similarly, there is accumulating evidence that rapid inflation, usually born of monetised budget deficits, impedes growth (e.g. Fischer, 1993) and worsens inequalities.

Hence, while there is fierce controversy about other aspects of their record, there appears to be general agreement that their efficiency in macroeconomic management has been a major reason for the remarkable growth of the East Asian 'Tiger' economies.[2] Similarly in Latin America, a region which traditionally placed macro management rather low in its pecking order of policy priorities, research indicates macroeconomic stability to be a major determinant of economic growth (De Gregorio, 1992). There has been a large turn-around in government attitudes to macro management in this region, so that what formerly was anathema has now become mainstream. Most of the heat has gone out of the controversies that used to rage there about the policies of the IMF. To the extent that they convey the message that macro management is of second-order importance, or that an excess of demand over supply can be eliminated painlessly, the critics of the Left are doing a disservice to those whose welfare they are trying to promote.

Confronted in these terms, the critics tend to respond that they are not against macro management, but against the particular approach to this of the Fund (together with the World Bank). Later in this chapter we shall consider specific ways in which the Fund approach might be improved, but ours are incremental suggestions in a reformist tradition which the more radical critics reject. To the limited extent that the critics go beyond negativism, they search for Another Way, an alternative mode of economic management which would combine stability, growth, greater equality and declining poverty. Unfortunately for them, it is no accident that orthodoxy today holds such sway, for orthodoxy itself has adapted (Chapter 2) and the search for radical alternatives has come to little.

The history of Latin America 'heterodox' programmes in the 1980s illustrates this point, for it was in that region, with its strong radical and structuralist traditions, that the search for alternatives was most vigorous. The chief of such programmes were the *Austral* and *Cruzado* plans of Argentina and Brazil respectively, both of 1985, and Mexico's *Pacto de Solidaridad* of 1987–8 – programmes which between them covered the largest countries of the region (see Kiguel and Livitian, 1992, and the sources cited by them).

These had three key features: they were addressed primarily to the control of inflation (rather than balance-of-payments deficits); the extent of their departure from orthodoxy was quite limited; and they failed to bring lasting superior results.[3] Essentially, they differed from the standard IMF-style model by including incomes policies, entailing price and wage controls – heterodox because non-market. But these incomes policies were in support of a standard demand-control package of tight fiscal and monetary policies. They certainly did not represent a thoroughgoing alternative to orthodoxy. Indeed, they have in a sense now entered the pantheon of orthodoxy, for it is now recognised that there are country circumstances (i.e. chronic high inflation) in which the temporary addition of price and income controls can increase the acceptability of anti-inflationary demand-reducing measures and weaken self-fulfilling expectations of renewed inflation.

Of course, there were also Latin American examples of more radical (or populist) attempts to escape the macroeconomic straitjacket, of which the policies of the Allende government in Chile (1970–73) or the Sandinista government in Nicaragua (1978–90) come to mind. However, whatever merits these governments may have had in other directions, they do not offer promising models for those seeking alternatives to the Fund, for each ended in economic crisis, with large public-sector deficits, hyperinflation and acute external difficulties. Indeed, their failure to manage their macroeconomies contributed to their political demise. The results of attempts elsewhere to devise some alternative, less painful, approach have been equally unpromising, e.g. Tanzania in the early 1980s, or Zambia in the late 1980s.

Lance Taylor has without doubt been the strongest intellectual influence on the 'heterodox' neo-structuralist school. But the conclusions of his latest writings reveal the limited extent of his disagreement with the 'Washington consensus' concerning macroeconomic management (1993: 87–8): sound macro policy is always desirable, but is not always feasible in developing country conditions and risks inducing secular stagnation; it is easy to exaggerate the gains from external liberalisation (over which the use of controls has some advantages) and privatisation; fiscal equilibrium is desirable but difficult to attain and to manage politically; changing real exchange and interest rates is difficult (but by implication desirable).

Loxley (1986) is another writer who sought overtly to develop, in the context of African-type economies, an alternative to Fund orthodoxies. Upon examination, however, his suggestions do not challenge conventional economic analysis in any fundamental way. They explicitly recognise the necessity for demand restraint, and are firmly in the reformist tradition (p. 141):[4] governments should be actively involved in programme design; programmes should give greater weight to the growth objective and to supply-side measures; explicit account should be taken of distributional effects, etc, etc. Indeed, as we saw in Chapter 2, Fund orthodoxy has gone some way to embracing such ideas.

The more sweeping criticisms of the Left are thus rejected as unsupported by the evidence, often based on a misattribution of responsibility, naïve and negative, having no radically different alternative to offer. What, now, of the Rightist critique, that the Fund is not orthodox enough, that it has mistakenly departed from the monetarist verities and assumed the role of a development agency for which it has neither mandate nor comparative advantage?

That the Fund *has* moved in such directions is undeniable. By comparison with the earlier decades of its existence and up to the time of the collapse of European communism, the Fund had come to be lending almost exclusively to developing countries. Even early in 1994, four-fifths of its programmes were in developing countries (making up 87 per cent of programme commitments, by value) and well over half were in low-income countries (Chapter 2, Table 2.2). Moreover, the SAF and ESAF accounted for over half of all outstanding programmes: medium-term programmes providing credit on highly concessional terms. Indeed, less than a fifth of its total commitments were in the form of traditional, short-term stand-by credits. Not only that, but the Fund approach is now a long way from a purely monetary one – concerning itself with institutional reform and supply-side policies in its EFF and SAF/ESAF programmes, even in some of its stand-bys. This was shown in Chapter 2, although we also showed there that it is easy to exaggerate the extent to which EFF and ESAF programmes depart from the Fund's traditional programme design.

What, in that case, might be said in response to the critique of the Right? First, that it is historically blind, in important respects. For one thing, the Fund never has been purely monetarist in its approach, so that to some extent the critics are harking back to a past that never existed. This was an issue explored in *The Quest* in 1984 (Chapters 5 and 6), with a conclusion which is worth repeating (pp. 219–20):

> it would be wrong to view the Fund as an organisation intent on the rigorous application of some theoretical model. In its attention to the current account [of the balance of payments] as well as the overall balance; its concern with fiscal matters generally ignored in the monetary approach; its attention to questions of domestic absorption and expenditure switching; its advocacy of devaluation as a policy instrument; the generally interventionist thrust of its programmes – in all these ways the Fund's practical work makes clear departures from a purely monetary approach.

However, while it never has been purely monetarist, it *is* true that in the 1960s and 1970s it confined itself more to short-term stand-by programmes built around the restraint of domestic credit, of the type regarded approvingly by the Rightist critics, than it does today. A second way in which these critics are historically blind, though, is that they overlook the fact that these programmes did not work well. This too was a subject explored in some depth in *The Quest* (Chapter 7), with the principal findings that stand-bys frequently broke down, had some tendency to be associated with short-term improvements in the balance of payments but with a low statistical significance, were not associated at all with sustained liberalisation, and tended to be associated with short-run *increases* in the inflation rate, although again with low significances.

A large part of the problem was diagnosed in *The Quest*, and by many others, as being due to the inappropriateness of stand-bys to the circumstances of developing countries, which already by the early 1980s were the Fund's principal customers. The limitations of this approach for developing countries are examined in detail in Chapter 4 above, where it is argued that heavy reliance on monetary policy is inappropriate in such economies, with the Fund itself quoted in support. In addition to the ever-present inadequacy of information about the state of the economy and how it works, it was pointed out that neither the income velocity of money nor the demand-for-money function is likely to have the degree of stability that application of the monetary model requires. Reasons are also given in Chapter 4 for doubting whether the monetary authorities have the control over the supply of money and credit they are presumed to have, and it is pointed out that their efforts to exert control sometimes impose inefficiencies on the economy, by way of credit misallocations. Moreover, the less developed the economy, and its financial system, the more acute these difficulties are.

Indeed, the Rightist critique can be regarded as a throw-back to an oversimplified Friedmanite rules-based approach to monetary policy which has never worked *even in industrial countries*, and is now abandoned in that form. To quote the Chief Economist of the Bank of England, 'No rule for monetary policy has been discovered which could credibly be followed' (The [London] *Independent*, 31 October 1994, p. 29). He is backed up by his Governor, who in the same source is quoted as adding 'The people to steer clear of in this field are those who pretend to know for sure . . . Even if we knew precisely where we were headed, we would not necessarily know precisely what to do with interest rates in order to get where we wanted to be.' They should know!

The Right is therefore advocating return to a monetarism of dubious value, and to programmes ill-designed for most of the countries that turn to the Fund for balance-of-payments assistance and which had a poor record in the period when they were the norm. It comes close to saying that the Fund has no role in low-income countries, with their more structurally based payments problems.

Some, not exclusively of the Right, have taken this line of thought further to suggest that it would be more appropriate for the Fund's 'structural adjustment' activities to be merged into the World Bank (see, for example, Commonwealth Study Group, 1983; Helleiner, 1992). The case for doing so is that the Bank is more experienced in this field, better equipped to deal with the structural weaknesses of low-income countries, more developmental in orientation, able to take a longer view, and has an Executive Board more sympathetic to low-income countries than its Fund counterpart.

Against this, we need to beware of giving the Bank more monopoly power. It is already very large, hard to change and by no means free of the traits of institutional arrogance. It too has its doctrinal hang-ups (not least a refusal to come to terms with the imperfections of market mechanisms in the conditions of low-income countries). It also has its gaps between policy rhetoric and the reality in the field, and is not always averse to steamrollering reluctant governments. We should also recognise the Fund's expertise in the design of macroeconomic and balance-of-payments policies, which the Bank would be unable to match. And we should remember that the survey of country experiences in Chapter 3 (pp. 106–9) found that Fund and Bank co-operation has been working quite well, through the mechanism of Policy Framework Papers and other means, even though there are inevitably disagreements in particular cases.

There is also the question, on what principle would IMF support be withdrawn from this category of its members? Formally at least, it is a co-operative, with membership open to all governments willing to subscribe to its overall objectives and with functions which are distinct from those of the Bank. Since the payments difficulties of many poor countries are acute, on what basis is responsibility for payments support for this class of countries to be transferred to another agency, except in the unlikely event that they would voluntarily agree? There would also be a distinct risk of a net reduction in the volume of concessional assistance available, attendant on a merger of the 'structural adjustment' lending programmes of the Fund and Bank, and we shall argue later that less finance is the opposite of what is needed.

There is a further aspect. Were its 'structural adjustment' (EFF, SAF, ESAF) activities to be merged into the Bank, the Fund would probably be left with few customers, chiefly some of the 'economies in transition' of Eastern Europe and the former Soviet Union (whose demand for the Fund's help is likely to diminish as their transitions proceed). Loss of these activities, on top of the much earlier loss of its roles as guarantor of exchange-rate stability and regulator of international liquidity, could deal the Fund a crippling blow. Those, on right and left, who would like to abolish it anyway might be well pleased. But we should remember an important difference between the Fund and the World Bank: unlike the Bank, the Fund (together with the GATT and its successor, the World Trade Organisation) institutionalises a set of rules to govern economic relationships between states and seeks to ensure that these rules are observed (Mayer, 1990: 151–2). In a world going through a nationalistic phase, and with great potential for economic instability, we should not lightly weaken one of our few mechanisms of international co-operation, for all its limitations.

The case for a merger is therefore unpersuasive, as is the view that Fifty Years is Enough. However, to say that there is a continuing role for the Fund in developing countries does not mean that there is no need for change. The weaknesses revealed in earlier chapters suggest the contrary, so we turn now to consider the desirable directions of reform.

## THE FUND'S LIMITED CAPACITY TO ADJUST

Implicit in our earlier response to the critics of the Left is the charge that they have failed to pay sufficient heed to the changes that have occurred within the Fund over the last decade or more, as detailed in Chapter 2. However, this case should not be overstated. As can be inferred from that chapter, the changes in programme design have been significant and in a desirable direction. Indeed, the extent to which the Fund has moved in the directions urged in *The Quest* is gratifying. But we should not get carried away: the change has been less than dramatic, in some cases quite modest, particularly in programmes in middle-income developing countries. Even ESAF programmes in low-income countries retain the hard core of the Fund's traditional conditionality. Indeed, these programmes are overtly presented as the most rigorous ('especially vigorous') of them all.

We should also not be unduly influenced by the change of tone from the top of the institution, from Managing Director de Larosière (1978–87) to Managing Director Camdessus (1987 to date), for there is a substantial gap between statements from the top and what occurs at the country level in the actual negotiation and execution of programmes. We observed this in Chapter 3 in the treatment of growth as an objective of ESAF programmes, and in the modest impact that the management's more sympathetic attitude has so far had in modifying programmes so as to lessen the risk of adding to poverty. While – because – there is no reason to doubt the genuineness of the desire to improve the growth and welfare effects of programmes, reducing the rhetoric-reality gap must be an important task for the future. Here the Fund could well emulate the World Bank in reporting regularly on the measures taken to implement these aspirations at the country level.

Chapter 4 shows that, despite the move towards longer-term and more structurally oriented programmes, doubts remain about the extent to which these have been adequately tailored to developing country circumstances, pointing out the potentially negative effects of high interest rates, or of premature trade liberalisation, or of cuts in public investment. It also suggests that the Fund has sometimes over-used devaluation as a policy weapon, with the danger of some of the ill effects of competitive devaluation which the Fund intended to avoid.

We have further drawn attention to the rather narrow view taken by the Fund (in co-operation with the World Bank) of what is meant by 'structural' in structural adjustment, with a large emphasis on improving price incentives, disengaging the state and strengthening the institutions of market efficiency, to the neglect of the more positive role that the state could play in the provision of economic services and infrastructure, industrial policy and export diversification (Chapter 2). This chapter has already stressed the unsuitability of reliance on monetary policy in countries with underdeveloped financial sectors. This is not only a serious intrinsic defect of the Fund's approach but leads on to what in Chapter 4 we describe as the indefensibility of the Fund's traditional performance criteria designed on the basis of its financial programming model and generating quantified budget and credit ceilings. That such criteria remain the chief vehicle for the Fund's conditionality is for no other reason, we suspect, than that nothing better has been found to put in their place.

## THE FUND'S OVER-RELIANCE ON CONDITIONALITY

That its key performance criteria are so lacking in intellectual credibility should raise questions about the Fund's reliance on conditionality as a means of achieving policy change, because performance criteria are the essence of its lending operations. And yet we have seen not merely that its modalities of conditionality have remained essentially unchanged over the years but also that conditionality has proliferated over the period studied, partly in consequence of the move into 'structural' measures. Chapter 2 showed that the number of performance criteria per programme increased more than 50 per cent during the 1970s and 1980s – a trend that may well have gone further since then – and that there was also an increase in the use of preconditions. Chapter 3, however, pointed out the difficulties the Fund has in securing implementation of its stipulations, with many programmes breaking down before the end of their intended lives because of non-compliance. It went on to make a link between this non-compliance and (with the chief exception of the exchange rate) the limited revealed ability of the Fund to exert a major influence on the policy instruments through which its programmes are intended to produce their results. It is scarcely surprising, then, that measurable programme effects are modest.

In fact, the proliferation of conditionality has intensified the non-compliance problem, which probably grows exponentially with the increase in the number of conditions. Moreover, it seems likely that non-compliance, in turn, distorts the content of programme measures in favour of those, like devaluation, which are easier to enforce, probably because they are amenable to treatment as 'prior actions' (preconditions). This may help explain the tendency to over-use devaluation.

There is, of course, no simple solution to the situation described above, but we see a change of strategy towards the negotiation of programmes as responding to a number of the difficulties. At present the Fund operates what can be called a 'pro-programme' strategy. By this is meant an approach which leads to programmes in a large number of countries, so that most governments which turn to the Fund with payments difficulties can expect to be able to agree a programme and secure financial assistance. This sounds desirable enough, but it has serious disadvantages.

One is that it encourages 'agreements' that exist mainly on paper. Such a development contributes to frequent programme breakdowns and a waste of scarce resources on countries whose governments are not serious about adjustment, or are incapable of delivering it. This, in turn, adds to the problem of arrears. The Fund's desire to safeguard past credits and avoid arrears adds to its incentives to make further loans to governments of dubious seriousness. More generally, the pressure to reach agreement results in the absurdity of doctoring projections in order to achieve a cosmetic improvement in forecast outcomes, and recourse to 'paper conditionality' which both sides know is unlikely to be honoured (Martin, 1991).

The number of programmes over-stretches the resources of the Fund. This leads to the under-funding of programmes, suggested in Chapter 3 as a source of failure, and reduces the Fund's ability to successfully pursue 'growth-oriented' programmes, which generally require longer time periods and more financing. The growth in the number of country programmes, combined with a US-led campaign to hold down the size of the IMF staff, has resulted in serious over-work and limited the time that can be devoted to any one programme. This reduces the ability to design programmes according to country circumstances, and strengthens institutional impulses towards application of a standardised approach, dealing with a rather narrow range of variables and uniform behavioural assumptions, of the type encapsulated in the financial programming model.

Thus, even though the policy thrust of the Fund is usually commendable, its programmes can get in the way by undermining the building of local capacities and responsive political systems which alone will permit adjustment to be sustained over time. It is perhaps the largest single cost of excessive reliance on conditionality that it undermines programme 'ownership', the importance of which we explored in the closing pages of Chapter 4. Conditionality poses a classical principal–agent problem, with differences in goals and interests between the principal (the IMF in this case) and agents (implementing governments); inadequate incentives for the agents to promote principals' objectives; asymmetrical information and high enforcement costs.

Apart from a laudable desire to offer assistance and an understandable sensitivity about, in effect, telling governments that it does not believe their promises, the Fund in its own defence could point to the political pressures brought upon it to agree programmes to favoured countries (Chapter 3). It has often not felt in a position to refuse assistance to governments, however sceptical it may have been about the seriousness of their commitment to macroeconomic prudence, unless it was willing to take on one or more of its major shareholders, and it has not always been resolute in the face of such pressures. However, the end of the Cold War has created a new situation, and it is already evident that geo-political considerations are impinging less on lending decisions. What is suggested, then, is that the IMF should take greater advantage of this new freedom to move from its pro-programme approach to a strategy of greater country selectivity, with an enhanced willingness to say 'No' in order to concentrate on helping 'serious' governments.

## THE CASE FOR GREATER SELECTIVITY

Perhaps the most powerful argument in favour of a greater selectivity, in which the Fund acts mainly in support of locally initiated programmes, is that such programmes have a greater revealed effectiveness. The most successfully adjusting group of countries are the East Asian 'miracle' countries (note, for example, South Korea's successful response to its large debt problems of the mid-1980s) *but their efforts owe little or nothing to Fund adjustment programmes.* Indeed, in important and well-known ways most of them departed from Fund (and Bank) orthodoxies, although both institutions provided valuable advice and support at various times. The same local ownership appears to characterise the current reform process in China and India. Similarly, it appears that the restoration of creditworthiness in the heavily indebted Latin American countries owes little to Fund programmes. Conversely, sub-Saharan Africa has undoubtedly been subjected to more conditionality per capita than any other region – and has achieved the least adjustment. Politically motivated changes in domestic government attitudes to macroeconomic management appear to be the decisive factor.

There are a number of reasons why we might expect locally initiated programmes to achieve superior results. For one thing, they are by definition designed around the characteristics of the domestic economy, tapping a local knowledge of this which international agencies are unlikely to possess. When discussing the neo-structuralist position in the preceding chapter we stressed the sensitivity of programme effects to the specifics of individual economies.

The design of macroeconomic policy is intrinsically highly political because it involves policies which affect the welfare of most people and can be expected to have quite large effects on the distribution of income, thus creating important groups of gainers and losers. Locally designed programmes can far more readily cope with such sensitivities. Being a product of domestic political and policy-formation processes, 'home-grown' programmes more faithfully reflect domestic goals and priorities. In the ideal case, the programme will be consensual, based on wide consultation and public information. Even in the absence of that, the government must be expected to have taken into account how the resulting social costs and political opposition are to be managed – something the Fund is not well placed to do. The probability of sustained government commitment to the chosen path of reform is enhanced.

The importance of this can scarcely be exaggerated. There is accumulating evidence that the nature of a country's polity – and the interventions that emanate from it – exerts a decisive influence on adjustment, for good or ill. Political systems have hampered it in sub-Saharan Africa, which has been marked by long-term government persistence with dysfunctional policies (Killick, 1995: Chapter 6). Conversely, there is wide agreement that governments in East Asia have been highly successful in promoting adaptation and experimenting with different policies, quick to drop those that have not worked and to try alternative measures (see World Bank, 1993: Chapter 4; Killick, 1995: Chapter 11). The substantial autonomy of policy-makers in these countries, i.e. their relative freedom from the influence of special-interest pressure groups, has been the central influence on the state's responsiveness and (tacitly) the resulting flexibility of the economy. Thus, Wade (1990) identifies the following factors to explain the success of Japan, Korea and Taiwan: centralisation of a decision-making structure employing the best managers; insulation of decision-makers from all but the strongest pressure groups; a powerful executive not beholden to the legislature; absence of a powerful labour movement; absence of conflicts between the owners of natural resources and manufacturers; and decision-makers' perception that their legitimacy is grounded in economic success.

If, as both logic and experience indicate, locally initiated programmes are more likely to be implemented and sustained than Washington-designed packages, this has the additional benefit of being more likely to call forth a positive response from the private sector. Because programmes will have greater credibility, the risk is reduced that they will be undermined by countervailing anticipatory actions by private agents. Being regarded as providing more reliable signals about the future, the probability is increased that domestic private investors will respond, and also that foreign capital will flow in (as it has in the successfully adjusting countries of Asia and Latin America), because foreign investment is sensitive to macroeconomic conditions (Schneider and Frey, 1985).

A shift by the Fund to a more selective support of locally initiated programmes should have the further merit of reallocating financial resources away from reluctant adjusters and client states. In short, a greater willingness to say 'No' should offer a better rationing device for scarce resources, increasing the effectiveness of their use and diminishing the risk of under-funding well-founded programmes.

In short, the IMF should recognise that its main contribution to successful adjustment in developing countries has been through its influence on the contemporary intellectual climate in which policy issues are debated, and persuasion of governments and their advisers through the regular contacts that occur. If the achievements of its programmes have been overstated, there has also (because it is hard to demonstrate) been an under-acknowledgement of this intellectual influence on the 'silent revolution' that has occurred in government attitudes towards macroeconomic policy. On this view, the turn-around in Latin America can still be viewed as a success story for the Fund (among others), but not one achieved principally through the specifics of its conditionality.

Home-grown programmes should not be over-sold, however. They do not guarantee success. The possibility of mis-design is still present and political and other shocks can supervene. One of the findings of Chapter 3 was that government–Fund consensus about programme content offered no guarantee of implementation in the face of domestic political opposition. Moreover, it is not our position that conditionality is *never* effective. It – and the money which comes with it – can be potent when it tips the balance between evenly poised domestic forces promoting and opposing policy reform, enabling vested interests to be bought off or confronted. In the general case, however, to be effective adjustment measures require an understanding on the part of responsible ministers of the actions necessary, with policies emerging organically, as it were, through local decision and implementation processes, and tailor-made to domestic conditions in a way that is only feasible when they are designed locally. Such programmes can be presumed to stand a better chance of success than Washington-designed programmes which are, to a substantial degree, wished upon more or less reluctant governments desperate for money.

Movement in the direction of support of local programmes would have major implications for the way the Fund currently conducts its lending. Recourse to conditionality would be much reduced, if not dropped altogether, and instead the Fund would need to seek ways of maximising its influence on policy-makers and their advisers. In this respect, the existing Article IV consultations are an excellent device. Another possibility is a greater decentralisation, with a larger number of permanent resident representatives who are endowed with greater authority than is currently the case. The Fund's research also has a crucial role to play, subject to two cautions: (a) it should avoid any tendency to confuse the reportage of research with propaganda, as seriously erosive of intellectual credibility; and (b) it should ensure that its researchers do not become estranged from field operations (as they have tended to be in the past). Of course, none of these actions could make all governments sweetly reasonable, but there is an incentive mechanism at work here: governments that make a mess of their countries' economies are apt to lose office.

Two further cautions. First, it is important not to confuse the limited technical capabilities of some developing country administrations with unwillingness to adjust. Technical assistance to enhance such capabilities should be freely available *but*: it should be independent of the Fund, so as to minimise conflict-of-interest problems; and it should *never* be imposed. Imposed advisers are no more effective than imposed policy reforms, as forcefully conveyed by the World Bank Vice-President responsible for Africa (Jaycox, 1993):

> Donors also relied too heavily on foreign experts, even when qualified Africans were available. This did little to foster a receptive environment for the transfer of skills. In fact, it was often bitterly resented. Over-reliance on technical assistance also brought many difficulties. Expatriates were frequently chosen for their technical skills rather than their ability to pass on those skills. This, coupled with operational difficulties, pulled foreign consultants into operational support at the expense of capacity building.

Second, insistence that programmes be domestically designed would require the Fund to be more pragmatic and pluralistic in its evaluations of the programmes submitted. It would be no escape from the ineffectiveness of conditionality if the only acceptable policy packages were identical to how the Fund staff would have written the Letters of Intent under the former modalities.

## OTHER REFORMS

Assuming for the moment that conditionality is retained, we should return to one of the themes of Chapters 3 and 4, about the treatment of growth in IMF programme design. We showed there that, despite strong statements from its Managing Director and an overt assertion that growth would be given greater priority in ESAF programmes, the reality even in those programmes, let alone stand-bys, is that not much has changed. The Fund continues to be unwilling to allow some minimum rate of economic growth (which might be set at, say, 1 per cent per annum above population growth) to act as a constraint upon the design of its demand-control measures. Instead, as we saw in Chapter 3 (pp. 83–5), ESAF programmes appear actually to have been associated with an economic slow-down when compared with the (lower-conditionality) SAF programmes which usually preceded them.

To introduce such a constraint would require substantial modification of its present approach to programme design and a move in the direction of the 'growth exercises' recommended in the Group of Twenty-Four's 1987 report (p. 16) (see pp. 149–50). In particular, it would require an approach to calculations on financing which incorporated estimates of the amounts needed for the minimum growth target to be met and a mechanism for finding the money, rather than starting, as at present, from an estimate of what finance is likely to be available. It would also be desirable to de-link the scale of the Fund's own credits from the size of borrowing countries' IMF quotas, and to allow credit size to be determined more by needs. There is already some scope for this, but Fund staff have to plead exceptional circumstances, which is not always feasible.

The suggested change in procedure would not be without difficulties, of course. The uncertainties involved in gap calculations would permit no better than rough estimates, and it would be hard to safeguard against the massaging of data in order to arrive at some predetermined figure. But Fund staff are used to working with estimates subject to large margins of error, as Chapter 4's account of the derivation of credit ceilings made clear, and there are limits to how much manipulation could be rationalised.

Another implication of taking the growth objective more seriously would be the necessity of safeguarding public investment during programme periods. We found in Chapter 3 that programmes are strongly associated with reduced investment levels, mostly as a result of cuts in governments' capital spending. This tends to be self-defeating: it is difficult to envisage any sustained combination of satisfactory growth and productive restructuring without substantial investments, particularly against the evidence that private sector investment is positively influenced by capital formation in the public sector, especially in infrastructure.[5]

Chapter 3's survey of the evidence on the impact of Fund programmes on poverty concluded that the effects were complex and varied considerably between countries, so that sweeping allegations to the effect that most adjustment costs were borne by the poor could not be sustained. However, there *is* evidence that some of the poor are especially vulnerable, particularly groups of the urban poor, and that, while it is not possible to demonstrate that programmes have worsened poverty overall, there is equally little evidence that they have improved it either. As reported in Chapter 2, the Fund now discusses with governments how vulnerable groups can be protected and some programmes do now contain safety-net provisions, especially for retrenched civil servants, but it is doubtful how much difference this has made in practice.

How might this situation be improved upon? Within the context of existing modalities, there should be a much more systematic institutionalisation of poverty considerations in programme designs and submissions to the Executive Board. Programme documents should include substantial coverage of likely poverty effects and programmes should not be approved which are likely to lead to serious impoverishment for which no compensatory provisions are included. In other words, the avoidance of serious impoverishment ought also to be accepted as a constraint on programme designs. The Managing Director (who has a personal interest in this) should publish a detailed annual report on progress made on the poverty alleviation front. If we go beyond the framework of existing modalities, in the direction of the more selective approach advocated above, then government willingness to safeguard its poor should be one of the criteria for selecting the programmes to be supported. Indeed, Managing Director Camdessus has hinted at the possibility of movement in that direction, suggesting that governments which take their social obligations seriously are more likely to attract international support than those which waste money on prestige projects and military spending (reported in Polak, 1991: 26).

The idea of a report on programme welfare effects introduces another item onto the agenda of IMF reform, namely the desirability of greater openness and transparency. A good deal of momentum in this direction has already been generated by the US Congress, partly in connection with controversies about Fund lending to Russia. Certainly, there is much scope for improvements in this direction, both in order that borrowing governments may know what the rules are and that the Fund can be properly monitored by shareholder country representatives who can adequately inform themselves about its policies and operations.

We need to make a distinction here between the, often highly sensitive, country-specific information which the Fund obtains in the course of programme negotiations, and routine country economic data, or data which are not country-specific. The Fund can reasonably protest that to place the former category into the public domain without the member government's consent would be improper and would undermine the frankness of programme negotiations. However, its secretiveness goes far beyond what can be justified in these terms, to border on the obsessive. The present writer has devoted a good deal of the past 15 years to undertaking research on the IMF, but even so there are large areas that he can only guess at, or can only try to deduce indirectly from intimate knowledge of Fund sources.

To take two examples: on pp. 58–66 of Chapter 3 there is an

analysis of programme breakdowns which adds usefully to information about the effectiveness of Fund programmes. But to analyse what happens to programmes involves a careful programme-by-programme tracking, including study of the footnotes of a particular table in the Fund's *Annual Report*. The Fund has responded to this analysis by pointing out, quite rightly, that not all programme breakdowns represent failures, but without being willing to provide the information that would be necessary for a more nuanced analysis. Why should such information not be available to representatives of its shareholders, leaving aside the special pleading of academic researchers? If that means being more frank about the performance of member governments, so be it: the World Bank is routinely far more forthcoming about its views on borrowing government policies.

A second example can be taken from the discussion in Chapter 4 of the Fund's use of waivers, which we describe as shrouded in mystery. Senior officers in the Fund assured the present writer that access to waivers is not used as a policy instrument, as 'a tap to be turned off or opened wide' as Chapter 4 puts it, but this was later contradicted by other information that the Fund had subsequently adopted a policy of granting waivers more readily in response to external shocks; in other words, it had opened the tap wider! Such a policy action should be in the public domain, so that shareholder representatives (and their constituents) know what is going on and borrowing governments know the rules which govern their transactions with the Fund.

A last topic to be taken up here concerns the governance of the Fund, specifically its continuing domination by the United States and the other shareholder governments who together make up the G-7 group. One of the advantages of the Fund (together with the World Bank) is that it has avoided the one-country–one-vote rule of the United Nations, with all the debilitating consequences that has had. Unfortunately, it has drifted towards an opposite and undemocratic extreme. While G-7 domination could be justified in the 1950s and 1960s in terms of the share of these countries in world trade and financial flows, such a defence is far harder to sustain today, with the growing economic importance of Asian and Latin American 'newly industrialised' countries. The United States retains a dominant voice even though its relative global economic importance has declined. By contrast, developing countries collectively have a smaller say in the Fund's councils than they had twenty years ago, even though they now account for nearly two-fifths of total world output,[6] have emerged as major exporters of manufactures, and host increasingly important international capital markets.

Some movement towards greater democratisation is long overdue, even if it has to be balanced pragmatically against the interests of those who provide most of the Fund's resources. Some saw signs of a revolt against the present position when, in October 1994, developing countries used one of the rare opportunities when their collective strength could tip the balance, to refuse a package on a new issuance of SDRs. The G-7 thought it had settled this as an internal compromise within the group, which would have left developing-country interests largely to one side. Both the past dominance of the G-7 and the potential significance of this event were well caught by the comment of the Chairman of the Interim Committee:

> There was a demonstration today that, in certain circumstances, all members cannot be forced to accept what the G-7 has decided the previous day. In a way, it is unprecedented. I have attended this meeting for some years now, and it was a little refreshing to see that all was not settled the previous day.[7]

## REDRESSING THE FINANCE–ADJUSTMENT IMBALANCE

Incorporated in the Bretton Woods settlement at the end of the Second World War was the idea that international finance was desirable for countries facing balance-of-payments difficulties – in two circumstances. It was desirable (i.e. in the interests of the wider international community), first, to encourage countries to ride out problems of a temporary, self-liquidating nature, e.g. caused by short-term commodity price slumps or harvest failures. Policy adjustment (cutting aggregate demand) was not regarded as appropriate in such cases, because it would impart a deflationary bias to the domestic and world economies, and because the time lags involved made it unlikely that the policy corrections could produce the needed results quickly enough. Second, when a balance-of-payments problem was not regarded as temporary, then domestic policy corrections were appropriate, and in this case financial support from the IMF (and other sources) was seen as complementary to adjustment, facilitating action, giving the deficit country more time to achieve results, and alleviating the adverse social impact of the policies.

Over the years, and particularly since the beginning of the 1980s, this philosophy has been systematically undermined. A number of the changes in the Fund reported in this volume have been symptomatic of this regression. One illustration is provided by the policies adopted towards heavily indebted countries after the debt crisis broke in 1982. These countries (not surprisingly) experienced a dramatic decline in access to world capital markets and the policies adopted by creditor governments were intended to keep any compensating increase in public financial provisions to an irreducible minimum, forcing many debtors to embark upon drastic deflationary policies. This was not an efficient response because the debt overhang acted as a tax on economic reforms (Sachs, 1989a): the country's population bore the cost of reforms while the creditors appropriated many of the benefits.

The change in philosophy is revealed too in the changes, reported in Chapter 2, that have occurred in the former Compensatory Finance Facility, which shifted from the semi-automatic provision of finance as an alternative to demand deflation in the face of temporary export shortfalls to a fund effectively only available to provide additional money for high-conditionality programmes. This prompted the Fund's *éminence grise*, Jacques Polak (1991: 3), to observe that the traditional distinction between conditional and non-conditional Fund credit 'has ceased to exist for all practical purposes'.

Less well known but even more starkly illustrative is the trend, also reported in Chapter 2, in the Fund's approach to contingency planning, with its programmes not only requiring the accumulation of additional reserves (by means of more stringent policies) against the possibility of adverse external movements but also requiring the introduction of tighter measures in the face of favourable shocks. On this logic both adverse and favourable shocks require tighter policies! Nothing could more clearly illustrate the movement away from internationally co-operative solutions than that the possibility of unforeseen developments should require anticipatory reductions in demand rather than international finance in the event of need. This illustration is, moreover, much more than theoretical. The survey reported in Chapter 3 (pp. 87–93) shows that 'exogenous shocks' are a common source of balance-of-payments difficulties, and other evidence presented there demonstrates a statistically strong association between adverse terms-of-trade movements and programme breakdowns.

The country studies reported there also reveal the frequency with which natural disasters, including droughts and hurricanes, are an important, sometimes dominant, factor in decisions to adopt an IMF programme. Stabilisation programmes cannot fail to be an inefficient response to such events. It contradicts the precepts of economic management that activity should have to be cut back as a result of such shocks. Countries which are particularly vulnerable to disasters – like Bangladesh, Africa's Sahelian states and many small island economies – require measures which will minimise the risks (e.g. through early-warning systems) and which cushion their economies against the adverse effects of these events when they do strike (such as the maintenance of stabilisation and contingency funds). The economies in question are often so poor and hard-pressed that they are limited in what they can do along these lines, which points to a need for international support. Although some assistance is available, it is neither large nor rapid enough to prevent an inherently inefficient policy response from becoming inescapable.

One factor which may have contributed to this turning away from the provision of public finance is what it is only a modest exaggeration to call *the myth of the Fund's catalytic effects*. It is widely and strongly held in Washington and among major shareholders that Fund programmes exert a powerful inducement effect on other sources of capital, thereby reducing the volume of credit which the Fund needs to provide directly to achieve programme goals. The evidence surveyed in Chapter 3 tells a different story, however. The conclusion arrived at there is that, although catalytic effects can be observed in specific cases, the evidence of systematic net additional inflows is mixed. There appears to be no general catalytic effect on private flows, although there may be an inducement of additional public grants, loans and debt relief. Indeed, if we believe that international capital markets are efficient, it would be surprising if there were strong inducement effects on private flows. We have seen Fund programmes not generally to have large effects on countries' economic fundamentals, and the very act of negotiating a programme may signal economic weaknesses of which the markets were previously unaware. We cannot overstress the limited and unreliable nature of programme catalytic effects – and the unwisdom, therefore, of basing policies on an opposite assumption.

As is well known, the Bretton Woods arrangements were unable to avoid a fundamental disproportion in the required policy responses of surplus and deficit countries. The trends just mentioned have deepened this imbalance, moving even the Fund's official historian to observe (de Vries, 1987: 284):

> there is an understandable perception of asymmetry between developing and industrial country members in that the conditionality applied to the use of the Fund's resources has significantly affected developing members, while surveillance under Article IV . . . seems to have had little practical effect on the large industrial members.

The prime movers in this erosion of the principles of international co-operation have been the governments of the major OECD countries; it mirrors the 'conservative revolution' which has recurred in many of these countries since the early 1980s. The IMF as an institution has been largely powerless in the face of these developments, although it has not been specially eloquent in defence of the system of which it was the linchpin. Our point here, however, is that the positive developments in the Fund's policies described earlier have to be set in the context of a less co-operative international system, so that gains on the swings have probably been more than offset by losses on the roundabouts.

One of the chief losses has been an inability on the part of the Fund to provide the volume of finance that would be necessary to give reality to the rhetoric of 'adjustment with growth'. This has had a number of adverse consequences. It has prevented growth in the borrowing countries, and has not only retarded the welfare of their populations but has had some growth-retarding effects in the rest of the world. Second, and less obviously, it has tended to be self-defeating. To the extent that a disproportionate share of the reduction in absorption in borrowing countries has fallen on investment, this has retarded the economic restructuring necessary for better balance-of-payments performance in the future – a theme also taken up by Graham Bird in the companion volume (Bird, 1995: Chapter 4). Thus, inadequate finance now leads to greater demand for BoP support in the future.

Further, a relative scarcity of finance has led the Fund towards a reliance on the potency of increasingly rigorous conditionality which we have shown to be misplaced. Although we do not have access to the information that would permit this to be tested, it is a safe working assumption that the more stringent the conditionality, the less the government's 'ownership' of the programme and the scantier the implementation of its provisions. The more weight a shortage of financing places upon conditionality, the less likely it is that programme stipulations will be able to deliver the necessary economic improvements. So the Fund gets into the situation, shown in Chapter 3 (pp. 94–101) to be common, of countries returning to it time after time, sometimes with little strengthening in their payments situations to show for its support.

How might the deteriorating balance between finance and adjustment be improved upon? Our starting point must be that the major industrial countries show absolutely no willingness to inject large new amounts of public capital into the international financial system. It is therefore a matter of finding ways around this constraint.

If we take first the treatment of contingencies, one possibility is to devise a self-financing international insurance fund, to be administered by the Fund, which would provide participating countries with assurance of access to help in the face of natural disasters, sudden world price collapses and other adverse shocks, with premiums designed so as to be largest in times of favourable shocks. In the meantime, or perhaps as a complementary measure, the now nearly moribund Compensatory and Contingencies Financing Facility should be liberalised, with access being restored to its former low-conditionality basis and with assistance provided on concessional terms for low-income borrowers. As it is a fairly short-term revolving fund, the net financial costs of such changes would be modest.

When it comes to the provision of more financing for growth-oriented programmes, however, there is no pretending that this could easily be found. However, the position may not be as intractable as it seems. For one thing, the fact that the Fund's Board was on the brink of agreeing a new SDR issuance in late 1994 demonstrates that reactivation of the SDR and its use as a means of allocating more purchasing power to low-income countries, the case for which is argued more fully by Graham Bird in the companion volume (1995: Chapter 4), may not be the impossibility that it seemed previously. There may also be mileage in efforts to obtain agreement that future general increases in Fund quotas should be made more automatic, e.g. by linking them to increases in the value of world trade or some other objective proxy of the potential demand for the Fund's resources.

Finally, we should note three factors which relieve the severity of the financing problem:

(a) Among developing countries, the most intractable problems have become concentrated on a group of countries which, although numerous, are small in economic terms, with the implication that the scale of their financing needs is limited.[8] It is in sub-Saharan Africa and certain other of the least-developed countries that the worst problems are concentrated.
(b) The central recommendation of this chapter, that the Fund should be more selective in the programmes it supports and less reliant on its own conditionality, would act as a more discriminating rationing device than its present 'pro-programme' approach.
(c) A high proportion of the financing needs of creditworthy developing countries is once again being met through various private flows, and in forms less hazardous than the 1970s upsurge in syndicated commercial bank lending. It seems reasonable to posit some reallocation of the Fund's resources away from countries which have access to, and can afford, the world's huge private capital markets.

Admittedly, we have left the considerable needs of the 'economies in transition' of Eastern and Central Europe and the Commonwealth of Independent States out of this reckoning – needs which the G-7 have a livelier self-interest in satisfying than those of the least developed. These needs are seen as unpredictable but potentially large, and there is some danger that Fund support of the richer transition economies will reduce resources available for developing countries. On the other hand, the Fund's Board has hinted at the possibility of responding with a special quota increase should lending to these countries begin to threaten the Fund's liquidity (see IMF, *Annual Report*, 1994: 23).

On balance, then, it seems that it would not require some improbably large new infusion of G-7 support to meet the urgent needs of non-creditworthy developing countries for genuine 'adjustment with growth' based upon borrower-initiated policy packages. Of course, there is always opposition to change. But the Fund, never the most flexible of institutions, will have to be allowed to continue to adapt if it is to retain its (already limited) relevance to the world economy in the late twentieth century. Fifty years of the Fund have not exhausted its potential, but another decade without further reforms might well erode its international constituency of support beyond recall.

# NOTES

## 1 STARTING POINTS

1 For the period 1988/89 to 1992/93 the shares of developing countries were 79 per cent by number of credits and 78 per cent by value (Systemic Transformation Facility excluded).

2 The eleven countries in question were Cambodia (SDR 39m); Liberia (351m); Panama (181m); Peru (625m); Sierra Leone (86m); Somalia (126m); Sudan (1,008m); Vietnam (101m); and Zambia (862m). Source: *Annual Report, 1991*, Tables 5 and 6.

3 At the end of the decade the dramatic breakdown of communist rule in Eastern Europe and subsequently in the then USSR threw up yet further major new challenges. We neglect them here, however, because our focus is on policies towards developing countries.

4 Adapted from *IMF Survey, Supplement,* October 1993, p. 2.

## 2 CONTINUITY AND CHANGE IN IMF PROGRAMME DESIGN, 1982–92

1 The 'crowding-in' argument, for which there is a good deal of supporting evidence, is that public sector infrastructure and other investments induce higher levels of private sector investments, by raising expected rates of return, in contrast to those who argue that public sector spending 'crowds-out' the private sector.

2 Remarks to the UN in July 1986, quoted by de Vries (1987: 241).

3 Remarks to UN, July 1990, quoted by Polak (1991: 19). See Polak, pp. 17–22, for a discussion of the evolution of the growth objective within the Fund.

4 Its formal policies notwithstanding, however, repeated use has a long history among developing country members. Goreux (1989: 148) states that six countries had had credit outstanding for 20 to 29 consecutive years, and 15 others for 14 to 19 years.

5 Thus of 22 SAF and ESAF programmes for which a full programme year was completed during July 1988 and December 1989, all included benchmarks and/or performance criteria relating to domes-

tic credit, credit to the government and/or the public sector, and external borrowing; eight contained similarly binding provisions relating to the budget balance (from an unpublished 1990 staff report).

6 This is similar to the estimate of more than half of all programmes contained in *The Quest*, pp. 193–5, for 1973–81.

7 Goldstein (1986: Table 5) provides details of the policy content of 17 stand-bys agreed in 1980, of which only four contained trade liberalisation provisions. The proportion with 1980 EFF programmes was five out of thirteen.

8 See especially Tanzi (1989) on which the following paragraphs are based.

9 The earlier Fund position, that it was up to governments to decide how to implement overall cuts, has always been fairly extensively honoured in the breach, however, so we are describing a shift of emphasis, rather than a complete turn-around. Thus, the Fund's Deputy Director of its Africa Department until 1987 wrote that, after negotiating an overall expenditure cut, a Fund mission 'has to discuss with the authorities how to implement this cut by analysing the budget line by line' (Goreux, 1989: 143). See also *The Quest*, p. 190.

10 Thus the number of IMF technical assistance missions is reported to have more than doubled between 1988/89 and 1992/93, from 270 to 606, with the number of person-years devoted to this nearly doubling over the same period. Source: *IMF Survey*, 30 May 1994, p. 170. This source provides a useful account of the Fund's technical assistance work.

11 To illustrate from one of the few country programmes which is publicly available, the stand-by approved for Cameroon in September 1988 included provisions for the preparation of an action plan for the forestry sector, movements towards the rationalisation of industrial protection, the rehabilitation and restructuring of the commercial banking system, and increased taxation of petroleum products. However, the all-important performance criteria were restricted to the conventional ceilings on domestic credit, credit to government and new external debt, and minimum reductions in external arrears.

12 The following is based on an unpublished 1991 IMF study of 28 SAF and ESAF programmes in 22 countries considered by its Executive Board by April 1991.

13 Stiles (1991: 198) also reports a substantial degree of IMF flexibility in his study of country programme decision processes.

14 Mosley *et al.*'s study of World Bank structural adjustment lending arrives at a similar conclusion (1991, Vol. 1: 125): 'significantly tighter conditions were negotiated with the poorer countries in the sample, those with the worst balance of payments problems, and those most dependent on [structural adjustment] finance for official capital flows from abroad. The poorer the recipient's initial *political* bargaining position, in other words, the more stringent the conditions

imposed upon it, regardless of the severity of *economic* mismanagement . . .'.

15 The overlap in the first two periods is presumably because the data are taken from staff studies which contained these overlapping periods.

## 3 PROGRAMME EFFECTS: WHAT CAN WE KNOW?

1 Although it is useful shorthand to refer to 'IMF programmes', the Fund always refers to 'Fund-supported' programmes in order to convey that the programmes are the property of the government which sign them. The extent to which this is truly the case, is of course, one of the points of controversy in the literature. Our shorthand use of 'IMF programmes' is not intended to prejudge this issue.

2 Khan and Knight (1988) tested for the influence of import availability on the export performance of 34 developing countries and found a large and highly significant positive correlation between them.

3 For a discussion of this and other reasons for programme 'failure' in the 1970s see *The Quest*, pp. 250–64.

4 For a more extended discussion of the methodological issues, although excessively dominated by concerns about the counterfactual, see especially Goldstein and Montiel (1986). Also Khan (1990).

5 A substantial number of researchers have used this method. Among the principal reports of research findings are: Reichmann and Stillson (1978), Connors (1979), *The Quest*, Goldstein and Montiel (1986), Khan (1990) and Nashashibi *et al.* (1992).

6 See Reichmann (1978); and Zulu and Nsouli (1985). The results of some unpublished IMF tests are reported in *The Quest*, (Chapter 7). See also Edwards (1989b). Heller *et al.* (1988) also contains target–actual information.

7 The only such example of which we are aware is provided by Edwards (1989b) who obtained information on programmes in 1983–85 – see his Table 5.

8 Thus, Gylfason (1987) went to considerable lengths to ensure that his programme and non-programme countries had BoP difficulties of equivalent severity. However, it is likely that some of his control group failed to go to the IMF on ideological grounds, or due to incompetence, or because they gave low priority to macroeconomic management. This group could thus be expected to do poorly in macroeconomic terms, so that their experiences were not a good proxy for what would have happened in the programme countries in the absence of the Fund. In general, the contrasts would impute unduly positive results to the programmes. Examination of the composition of Gylfason's control group reveals this to be very heterogeneous, with large dispersions around the mean values of the variables used by him.

9 I have borrowed this expression from Khan (1990). There have been various uses of this method in single-country case studies. The chief

examples of use in cross-section analyses are Khan and Knight (1981 and 1985), and Doroodian (1993).

10 Again, I have borrowed from Khan (1990) which is also the chief example of the use of this technique. See also Goldstein and Montiel (1986).

11 This technique was used in an earlier ODI study, with the case studies reported in Killick (1984b) whose results are utilised in the overview *The Quest.* The WIDER country studies, whose results are summarised in Taylor (1988), also use this method, although this project does not attempt to measure the effects of IMF programmes *per se.* See also Heller *et al.* (1988).

12 It is difficult to judge what is the best way of proceeding on this. As is shown in the discussion of experiences with ESAF (pp. 76–85), growth has increased in prominence as a stated objective of IMF programmes in recent years, particularly in ESAF programmes, but the growth objective remains clearly subordinate to BoP goals, even in ESAF programmes. For present purposes, it makes little difference whether we classify growth as an objective or a by-product.

13 A partial exception is in *The Quest* (pp. 240–2), which on 1970s data found a significant initial reduction but that this was not sustained into the second year.

14 Throughout this chapter Year 0 refers to the twelve months (or the calendar year in some cases) immediately following adoption of a Fund programme.

15 His and other results reported earlier are a weighted average for 1983–5 calculated from the information in his Table 5.

16 See especially the studies sponsored by UNICEF in Cornia *et al.* (1987 and 1988). These are critical of the effects of specific programmes and of the past neglect of distributional aspects in the design of IMF programmes but are guarded in their general assessments of programme effects.

17 An official listing of (usually publicly available) IMF Working Papers includes an April 1989 paper by Joshua Greene on the effects of Fund programmes in Africa, 1973–86, but the Fund refused to release this. It may be conjectured that this is because the study found unacceptably weak programme effects in these countries.

18 The following paragraphs have benefited from access to Dane Rowlands's (1994) study of the response of new lending to countries' agreement on an IMF programme.

19 In his report on the WIDER study, Taylor (1988: 144) similarly emphasises the difficulties of making any simple statements about catalytic effects. Connections between capital flows and macro stabilisation efforts are complex, he suggests, especially given the turbulence in the 1980s in bank financing and debt servicing.

20 See also Özler (1993) who reports for 1955–68 that countries adopting Fund stand-bys were subsequently charged higher interest rates on new loans by private creditors.

21 I should like to acknowledge the major contribution of Moazzam Malik to this work. It was he who undertook the detective work on

programme breakdown and subjected the results to statistical analysis. In turn, Ramani Gunatilaka did an excellent job of up-dating the earlier work for the purposes of this book. The chief sources from which the relevant information was gleaned were various issues of the IMF's *Annual Reports*, augmented by issues of the *IMF Survey.*

22 As reported later, it appears that a high proportion of ESAF programmes also do not remain on track, in that they need to be extended beyond their intended period of implementation. The position with SAF programmes is different again, for the level of conditionality attached to these programmes has been rather nominal and there is hence little reason for expecting high breakdown rates for them.

23 The following comparison of results is based on the original tests, reported in Killick *et al.* (1992a, 1992b). The non-completion tests have been updated for the purposes of this book but there is no reason to think that their reliability has diminished in the meantime.

24 The test we applied to ascertain this was that less than 25 per cent of the credit was utilised.

25 The values used in the compilation of Table 3.4 are annualised averages of actual uses ('purchases' in the Fund's parlance) of credits during periods when the programmes in question were operative. In cases where programmes were suspended or abandoned, only purchases during the period when the programmes were active have been included and the period adjusted accordingly when annualising the data. The base value is taken as the mean annualised value of the variable in question during the two years immediately preceding the programme.

26 For programmes begun in the period 1983–5 there was a mean reduction in import volumes of nearly 9 per cent on the average values in the immediately following two years. Using a one-tailed test, this was significant at the 95 per cent level. It should be added, however, that there was a considerable spread of values around the means.

27 One difficulty here relates to poor data and the variety of definitions for the overall balance. The results presented here refer to the change in the official foreign reserves. Using the *International Financial Statistics* performance balance definition yields an even smaller and insignificant improvement.

28 The apparently statistically significant doubling of direct foreign investment unfortunately only illustrates the dangers of relying on t-statistics when working with a small sample. If the two programmes agreed with Dominica are excluded the increase becomes negligible. A similar explanation accounts for the bizarre result that official transfers *decrease* for completed programmes but increase for uncompleted programmes. This result is entirely the result of large changes occurring just in Dominica. Excluding Dominica reveals that for the rest of the sample there was a significant increase in official transfers, increasing by 30 per cent on average four years after an agreement and increasing to some extent for three-quarters of all programmes.

29 The substance of this section is derived from a forthcoming review article in *The World Economy* (1995) and is reproduced with permission. See also the 'Reply' by Schadler in the same issue.

30 The test yielded only t = −0.85, against a minimum value at the 90 per cent level of ±1.75 and 2.12 at the 95 per cent level.

31 All four countries named in *OP 106* as strong adjusters (Bolivia, Gambia, Ghana and Guyana) are classified as countries having made relatively good progress towards external viability; and all five countries named in the unpublished (IMF, 1993) version as weak adjusters are included among those making little progress towards viability.

32 This is consistent with the results of recent econometric work by Easterly *et al.* (1993), concluding that 'Shocks, especially those to terms of trade, play a much larger role in explaining variance in growth rates than previously acknowledged' (p. 1).

33 The report, Table A1, shows a mean population growth rate for the 19 countries of 2.7 per cent and a median of 2.8 per cent.

34 These issues are surveyed, mainly with reference to the 1970s, by Killick and Sharpley in *The Quest* (Chapter 2). See also Khan and Knight (1982).

35 Other countries where we judge that their decision to request a Fund programme was the result of a combination of exogenous shocks and domestic policy weaknesses are: Côte d'Ivoire, the Dominican Republic, Malawi, Morocco and Tanzania.

36 Note that it is possible for more than one programme to be in operation at any one time, e.g. when a stand-by and a SAF programme run concurrently. This explains the surprisingly high ratio of programmes to years in operation. In addition to these eight countries, the Philippines, Somalia and Sudan were others from our sample which had frequent recourse to the Fund. However, they are not included in the following analysis because of the early breakdown of many of the programmes in these countries, and poor information.

37 The reader is cautioned, however, that the analysis is largely based on scrutiny of a range of statistical indicators which sometimes differ widely in alternative official sources and appear sometimes to be subject to large error margins. It is therefore difficult to be as confident as we would like to be about the judgements offered.

38 The statistics also show an improvement on the current account, but the figures here are strongly influenced by increased levels of official transfer payments (i.e. grant aid), which are probably best regarded as capital transactions. If official transfers are excluded, the current account improves from deficit levels equal to nearly half of GDP in 1980–81 to 10 per cent in 1988, although there was a sharp increase again in 1989.

39 We might also mention that the World Bank regards weak macro management policies as an important cause of failure in those of its SAPs which break down or are unsuccessful (World Bank, 1989).

40 This refers to the commonly observed tendency for per capita levels of aid to be inversely and strongly correlated with country size, as measured by population.

41 This argument is set out more fully in Killick (forthcoming) and is the subject of ongoing research into conditionality as a means of effecting policy change.

## 4 ISSUES IN THE DESIGN OF IMF PROGRAMMES

1 See IMF (1987) for an authoritative exposition and discussion of this model, from which I have borrowed heavily.
2 See IMF (1987: 14). See also Williamson's (1980) interpretation of the monetary approach to the BoP from a Keynesian perspective.
3 One illustration of this is provided by Khan and Knight (1985: 21–3) who employ a model based on IMF-type demand-management policies and find that, while the long-run growth rate is little affected, there is a substantial initial reduction. They also run a variant of the standard model which combines demand-management and supply-side policies, again finding a substantial short-term reduction in growth, but in this case a sustained longer-term improvement.
4 See also Khan and Knight (1982) who show for the 1970s that the payments problems of developing countries were chiefly, but not wholly, caused by deteriorating terms of trade.
5 In our study of the Kenyan case, Killick and Mwega (1993: 56) found an average lag of 18 months in the adjustment of the demand for money.
6 See Mundell, in Frenkel and Goldstein (1991: 482), on the static nature of the monetary theory of the balance of payments and the thus far only rudimentary attempts to dynamise it.
7 Support for this proposition is provided by members of the Fund's own research staff. See Khan and Knight (1988) who found, for a sample of 34 developing countries, a large and highly significant positive correlation between export volumes and the availability of imports.
8 See Aghevli *et al.* (1991) for a discussion for these and related issues by members of the Fund's Research Department. They point out that, as developing countries have moved towards greater exchange-rate flexibility, industrial countries have moved in the opposite direction.
9 See also the influential theoretical analysis by Lizondo and Montiel (1989) which concludes that 'the direction of the impact effects of devaluation on real output is ambiguous on analytical grounds' (p.182).
10 The source here is an unpublished Fund review of experiences with stand-by programmes in the period 1983–7.
11 From a 1991 review of experiences with ESAF programmes.
12 See Frey *et al.* (1984) for evidence of only a limited professional consensus concerning some basic propositions of monetarism even in the early-1980s. Cited by Spraos (1986: Table 3).
13 To give one counter example, Killick and Mwega's (1993) study of monetary policy in Kenya shows that the control of the Central Bank of Kenya over commercial bank lending is highly imperfect because

of delayed and unpredicatable bank responses to variations in their liquidity; that the central government has great difficulty in arriving at realistic estimates of its own deficit-financing requirements for the current fiscal year; and that the growth of near-bank substitutes ('non-bank financial institutions') has, in any case, substantially weakened the authorities' leverage over aggregate credit and demand in the economy.

14 See Tanzi (1989: 14) and references cited there. See also Mansur (1989) for the development and testing of what can be described as a fiscal model of the BoP, applied to the Philippines, finding a significant positive correlation between the fiscal and BoP balances, with causation apparently running from the former to the latter: the larger the budget deficit, the weaker the current account of the BoP.

15 See especially Tanzi (1989) from which the following exposition has largely been derived.

16 Martin and Mistry's survey of the preparation of Fund programmes in African countries amply demonstrates the frequent spuriousness of apparently objective payments and other projections (1991: 107):

> Projections . . . often underestimated the amount of external finance or debt relief creditors would provide, forcing lower current account deficit targets than necessary. Sometimes they overestimated external finance or underestimated debt service due, necessitating revisions to projections during IMF programmes. Occasionally they projected an optimistic picture . . . to convince creditors that they were not backing a 'basket-case' this often perversely produced less external finance than hoped . . . Crucially, little of the negotiation or last-minute juggling was matched by changes in conditions.

See also Martin (1991, Chapter 2).

17 For a further critique see Spraos (1986). He goes further than this to argue against placing target values upon policy instruments, as compared with target values for the BoP and other objectives being pursued. It is an incisive critique which, however, goes astray when he turns to propose an alternative, for he then quickly runs foul of the difficulty of reducing the BoP objective to a single quantified indicator, and of bringing in other goals.

18 See *The Quest* (p. 212) for documentation on how government access to waivers was consciously reduced in the early 1980s as part of a general toughening-up of conditionality.

19 See Martin (1991, Chapter 2) for evidence. The uncertainties just referred to can be linked to a wider-ranging argument developed by Rodrik (1991), who draws attention to the effect of policy changes initiated as a result of the adoption of adjustment programmes in increasing investor uncertainties. He suggests that even moderate uncertainty about policy acts as a substantial tax on investment, and that 'reform packages which emphasize policy stability and sustainability are likely to bring greater payoffs in terms of investment and growth than those which focus on economic liberalization and getting prices right' (pp. 240–1).

20 According to a senior World Bank staff member, during the course of negotiations with Malawi, Bank and Fund staff presented the government with two alternative sets of projections, one based on a growth target and the resulting financing needs and the second based on the likely available finance. Reportedly, however, this caused so much confusion on the Boards of the two institutions that 'we will never do that again'.

21 See World Bank (1992: Chapter 10 and Annex 8), and also Johnson and Wasty (1993).

22 See, for example, Nelson (1989 and 1990), who emphasises the importance of the quality of political leadership and writes of the existence of a 'reform syndrome' as presenting the best circumstances for effective adjustment programmes: 'Leaders firmly committed to major change, widespread public acceptance of demand for such change, new governments with strong centralized authority, and a disabled opposition constituted the political context for determined adjustment efforts' (1989: 12).

23 It is also significant that the positive assessment of PFP procedures in the Bank's evaluation of its structural adjustment programmes already cited (World Bank, 1992: 206–8) is largely couched in terms of its effectiveness as a way of co-ordinating the activities of the Bank and Fund.

24 Recall the conclusion of Chapter 3 that situations in which programmes are dictated by the Fund to a recalcitrant but desperate government are untypical. However, the absence of an adversarial relationship is not necessarily sufficient to ensure that the government will regard itself as owning the programme, for the chapter went on to point out that implementation was often poor even when there was broad agreement between the two parties.

## 5 CONCLUSION: IS FIFTY YEARS ENOUGH?

1 Based on reports by the InterPress Service of 18 January and 27 April 1994 of statements by OXFAM's senior economics spokesman and its Director, respectively.

2 See World Bank (1993: 121–3). Although this report on the East Asian 'miracle' economies has come in for much criticism, there has been no substantial questioning of its conclusions with respect to the importance of macroeconomic stability. Compare, for example, with Fishlow *et al.* (1994).

3 In Brazil and Mexico the programmes were followed fairly soon by hyperinflations. The Argentinian experience was more favourable, but this seems to have owed more to orthodox demand-control measures, e.g. to reduce the budget deficit.

4 For a similar set of proposals which I put forward with colleagues in 1984, see *The Quest*, Chapter 8.

5 Serven and Solimano (1993: Chapter 6) find that private investment is

positively associated with public investment, a result also obtained by Blejer and Chu (1989: Chapter 1) and Musalem (1989).

6 The 1992 proportion was 37 per cent, based on internationally accepted purchasing-power parity valuations of national accounts.

7 Philippe Maystadt, reported in *IMF Survey*, 17 October 1994, p. 316.

8 For example, the group of countries classified by the World Bank as 'severely indebted low income countries' in 1992 accounted for just over 12 per cent of total developing-country external indebtedness and a little over 1 per cent of total world exports. The whole of sub-Saharan Africa contributed about 1.5 per cent of total world exports and a little over 1 per cent of total world GDP. The whole of that region's GDP adds up to about one-half of that of the UK.

# BIBLIOGRAPHY

Aghevli, Bijan B., Khan, Mohsin S. and Montiel, Peter J. (1991) *Exchange Rate Policy in Developing Countries*, IMF Occasional Paper No. 78. Washington, DC: International Monetary Fund, March.

Bevan, D.L., Bigsten, A., Collier, P. and Gunning, J.W. (1986) *East African Lessons on Economic Liberalisation*. London: Trade Policy Research Centre.

Bird, Graham (1984) 'Balance of Payments Policy', in Tony Killick (ed.) *The Quest for Economic Stabilisation: the IMF and the Third World*. London: Gower and Overseas Development Institute.

Bird, Graham (1995) *IMF Lending to Developing Countries: Issues and Evidence*. London and New York: Routledge and Overseas Development Institute.

Blejer, Mario I. and Chu, Ke-Young (eds) (1989) *Fiscal Policy, Stabilization and Growth in Developing Countries*. Washington, DC: International Monetary Fund.

Bourguignon, F. and Morrisson, C. (1992) *Adjustment and Equity in Developing Countries: A New Approach*. Paris: OECD.

Brown, Richard P.C. (1990) *Sudan's Debt Crisis*. The Hague: Institute of Social Studies.

Brown, Richard P.C. (1992) 'The IMF and Paris Club Rescheduling: A Conflicting Role?', *Journal of International Development* 4(3), May–June.

Buiter, Willem H. (1985) 'A Guide to Public Sector Debts and Deficits', *Economic Policy* 1(1), November.

Cashel-Cordo, Peter and Craig, Steven G. (1990) 'The Public Sector Impact of International Resource Transfers', *Journal of Development Economics* 32.

Cleaver, Kevin (1985) 'The Impact of Price and Exchange Rate Policies on Agriculture in Sub-Saharan Africa', *Staff Working Paper* 728. Washington, DC: World Bank.

Commonwealth Study Group (1983) *Towards a New Bretton Woods* (The 'Helleiner Report'). London: Commonwealth Secretariat.

Connors, Thomas A. (1979) 'The Apparent Effects of Recent IMF Stabilization Programs', *International Finance Discussion Paper* No. 135. Washington, DC: Federal Reserve, April.

Corbo, Vittorio and Rojas, Patricio (1991) 'World Bank Supported Adjustment Programs: Country Performance and Effectiveness', *PRE Working Paper* No. WPS 623. Washington, DC: World Bank, March.

Corbo, Vittorio and Webb, Steven B. (1991) 'Adjustment Lending and the Restoration of Sustainable Growth', *Journal of International Development* 3(2), April.

Cornia, G.A., Jolly, R. and Stewart, Frances (eds) (1987) *Adjustment with a Human Face: Protecting the Vulnerable and Promoting Growth.* Oxford: Clarendon Press.

Cornia, G.A., Jolly, R. and Stewart, Frances (eds) (1988) *Adjustment with a Human Face: Ten Country Case Studies.* Oxford: Clarendon Press and UNICEF.

Cornia, A. and Stewart, F. (1990) 'The Fiscal System, Adjustment and the Poor', *Development Studies Working Paper* No. 29. Turin: Centro Studi Luca d'Agliano.

De Gregorio, José (1992) 'Economic Growth in Latin America', *Journal of Development Economics* 39(1), July.

de Vries, Margaret Garritsen (1987) *Balance of Payments Adjustment, 1945 to 1986: The IMF Experience.* Washington, DC: International Monetary Fund.

Dell, Sidney (1982) 'Stabilisation: The Political Economy of Over-kill', *World Development* 10(8), August.

Diakosavvas, Dimitris and Kirkpatrick, Colin (1990) 'Exchange-rate Policy and Agricultural Exports Performance in Sub-Saharan Africa', *Development Policy Review* 8(1), March.

Donovan, Donal J. (1982) 'Macroeconomic Performance and Adjustment under Fund-supported Programmes: The Experience of the Seventies', *IMF Staff Papers* 29(2), June.

Doroodian, Khosrow (1993) 'Macroeconomic Performance and Adjustment under Policies Commonly Supported by the International Monetary Fund', *Economic Development and Cultural Change* 41(4), July.

Easterly, W., Kremer, M., Prichard, L. and Summers, L. (1993) 'Good Policy or Good Luck? Country Growth Performance and Temporary Shocks', *Journal of Monetary Economics*, December.

Edwards, Sebastian (1989a) 'Exchange Rate Misalignment in Developing Countries', *World Bank Research Observer* 4(1), January.

Edwards, Sebastian (1989b) 'The IMF and the Developing Countries: A Critical Evaluation', *Carnegie–Rochester Conference Series on Public Policy* No. 31. Washington, DC: North–Holland.

Faini, Riccardo and De Melo, Jaime (1990) 'Adjustment, Investment and the Real Exchange Rate in Developing Countries', *Working Paper* WPS 473. Washington, DC: World Bank, August.

Faini, R., De Melo, J., Senhadji-Semlal, A. and Stanton, J. (1991) 'Macro Performance under Adjustment Lending', in Vinod Thomas *et al.* (eds) *Restructuring Economies in Distress: Policy Reform and the World Bank.* Oxford: Oxford University Press.

Feinberg, Richard E. (1991) 'The Bretton Woods Agencies and Sub-Saharan Africa in the 1990s: Facing the Tough Questions', in Ishrat

Husain and John Underwood (eds) *African External Finance in the 1990s.* Washington, DC: World Bank.

Fischer, Stanley (1993) 'Does Macroeconomic Policy Matter? Evidence from Developing Countries', *Occasional Paper* No. 27. San Francisco, Calif.: International Center for Economic Growth.

Fishlow, A., Gwin, C., Haggard, S., Rodrik, D. and Wade, R. (1994) *Miracle or Design? Lessons from the East Asian Experience.* Washington, DC: Overseas Development Council.

Frenkel, Jacob A. and Goldstein, Morris (eds) (1991) *International Financial Policy: Essays in Honor of J.J. Polak.* Washington, DC: International Monetary Fund.

Frenkel, Jacob A. and Johnson, Harry G. (1976) *The Monetary Approach to the Balance of Payments.* London: Allen and Unwin.

Frey, Bruno S., Pommerehne, W.W., Schneider, F. and Gilbert, G. (1984) 'Consensus and Dissension among Economists: An Empirical Enquiry', *American Economic Review* 74, December.

Garcia, Jorge G. and Llamas, Gabriel M. (1988) *Coffee Boom, Government Expenditure and Agricultural Prices: the Colombian Experience.* Washington, DC: International Food Policy Research Institute, August.

Goldstein, Morris (1986) *The Global Effects of Fund-supported Adjustment Programs,* IMF Occasional Paper No. 42. Washington, DC: International Monetary Fund, March.

Goldstein, Morris and Montiel, Peter (1986) 'Evaluating Fund Stabilisation Programs with Multicountry Data: Some Methodological Pitfalls', *IMF Staff Papers* 33(2), June.

Goreux, Louis M. (1989) 'The Fund and the Low-Income Countries', in Catherine Gwin and Richard Feinberg (eds) *The International Monetary Fund in a Multipolar World: Pulling Together,* US–Third World Policy Perspectives No. 13. Washington, DC: Overseas Development Council.

Ground, Richard Lynn (1984) 'Orthodox Adjustment Programmes in Latin America: A Critical Look at the Policies of the IMF', *CEPAL Review* 23, August.

Group of Twenty-Four (G-24) (1987) *The Role of the IMF in Adjustment with Growth,* Report of a Working Group. Washington, DC: Group of 24, March.

Guitian, Manuel (1981) *Fund Conditionality: Evolution of Principles and Practices.* Washington, DC: International Monetary Fund Pamphlet No. 38.

Gwin, Catherine and Feinberg, Richard (eds) (1989) *The International Monetary Fund in a Multipolar World: Pulling Together,* US–Third World Policy Perspectives No. 13. Washington, DC: Overseas Development Council.

Gylfason, Thorvaldur (1987) 'Credit Policy and Economic Activity in Developing Countries with IMF Stabilisation Programs', *Princeton Studies in International Finance* No. 60. Princeton, N.J.: Princeton University, August.

Hajivassiliou, V.A. (1987) 'The External Debt Repayment Problems of

LDCs: An Econometric Model Based on Panel Data', *Journal of Econometrics* 36.

Harrigan, J. (1991) 'Jamaica', in Paul Mosley *et al.* (eds) *Aid and Power: The World Bank and Policy-Based Lending*, Vol. 2. London: Routledge.

Harris, G. and Kusi, N. (1992) 'The Impact of the IMF on Government Expenditures: A Study of African LDCs', *Journal of International Development* 4(1), January.

Healey, John and Page, Sheila (1993) 'The Use of Monetary Policy', in Sheila Page (ed.) *Monetary Policy in Developing Countries.* London and New York: Routledge.

Helleiner, Gerald K. (ed.) (1986) *Africa and the International Monetary Fund.* Washington, DC: International Monetary Fund.

Helleiner, Gerald K. (1992) 'The IMF, the World Bank and Africa's Adjustment and External Debt Problems: An Unofficial View', *World Development* 20(6), June.

Heller, P.S., Bovenberg, A. L., Catsambas, T., Chu, K.-Y. and Shome, P. (1988) *The Implications of Fund-supported Adjustment Programs for Poverty*, IMF Occasional Paper No. 58. Washington, DC: International Monetary Fund, May.

Henderson, Anne (1992) 'The International Monetary Fund and the Dilemmas of Adjustment in Eastern Europe: Lessons from the 1980s and Prospects for the 1990s', *Journal of International Development* 4(3), May–June.

Hicks, N. (1991) 'Expenditure Reductions in Developing Countries Revisited', *Journal of International Development* 3(1), January.

Husain, Ishrat and Underwood, John (eds) (1991) *African External Finance in the 1990s.* Washington, DC: World Bank.

International Monetary Fund (1987) *Theoretical Aspects of the Design of Fund-Supported Adjustment Programs*, IMF Occasional Paper No. 55. Washington, DC: International Monetary Fund, September.

International Monetary Fund (1993) 'Review of Experience under ESAF-supported Arrangements'. Washington, DC: IMF, February (mimeo).

Jaeger, William (1991) 'The Impact of Policy in African Agriculture', *Working Paper* WPS 640. Washington, DC: World Bank, March.

Jaycox, Edward V.K. (1993) 'Capacity-building: The Missing Link in African Development'. Reston, Va.: African–American Institute, May.

Johnson, John H. and Wasty, Sulaiman S. (1993) 'Borrower Ownership of Adjustment Programs and the Political Economy of Reform', *World Bank Discussion Paper* 199. Washington, DC: World Bank.

Kafka, Alexandre (1991) 'Some IMF problems after the Committee of XX', in Jacob A. Frenkel and Morris Goldstein (eds) *International Financial Policy: Essays in Honor of J.J. Polak.* Washington, DC: International Monetary Fund.

Kamin, Steven B. (1988) 'Devaluation, External Balance and Macroeconomic Performance: A Look at the Numbers', *Princeton Studies in International Finance* No. 62. Princeton, N.J.: Princeton University Press, August.

Khan, Mohsin (1990) 'The Macroeconomic Effects of Fund-supported Adjustment Programs', *IMF Staff Papers* 37(2), June.

Khan, Mohsin and Knight, Malcolm D. (1981) 'Stabilisation Programs in Developing Countries: A Formal Framework', *IMF Staff Papers* 28(1), March.

Khan, Mohsin and Knight, Malcolm D. (1982) 'Some Theoretical and Empirical Issues Relating to Economic Stabilisation in Developing Countries', *World Development* 10(9), September.

Khan, Mohsin and Knight, Malcolm D. (1985) *Fund-Supported Adjustment Programs and Economic Growth*, IMF Occasional Paper No. 41. Washington, DC: International Monetary Fund, November.

Khan, Mohsin and Knight, Malcolm D. (1988) 'Import Compression and Export Performance in Developing Countries', *Review of Economics and Statistics*, May.

Khan, Mohsin and Montiel, Peter J. (1989) 'A Framework for Growth-oriented Adjustment Programs', *IMF Staff Papers* 36(2), June.

Khan, Mohsin, Montiel, Peter and Haque, Nadeem U. (1990) 'Adjustment with Growth: Relating the Analytical Approaches of the IMF and the World Bank', *Journal of Development Economics* 32.

Kiguel, Miguel A. and Livitian, Nissan (1992) 'When Do Heterodox Stabilization Programs Work? Lessons from Experience', *World Bank Research Observer* 7(1), January.

Killick, Tony (ed.) (1984a) *The Quest for Economic Stabilisation: the IMF and the Third World*. London: Gower Publishing Company and Overseas Development Institute.

Killick, Tony (ed.) (1984b) *The IMF and Stabilisation: Developing Country Experiences*. London: Gower Publishing Company and Overseas Development Institute.

Killick, Tony (1989) 'Economic Development and the Adaptive Economy', *Working Paper* No. 31. London: Overseas Development Institute, November.

Killick, Tony (1994) 'Structural Adjustment and Poverty Alleviation: An Interpretative Study', Geneva: UNCTAD (mimeo).

Killick, Tony (ed.) (1995) *The Flexible Economy: Causes and Consequences of the Adaptability of National Economies*. London and New York: Routledge and Overseas Development Institute.

Killick, Tony (forthcoming) 'Conditionality and the Adjustment–Development Connection', in IMF/World Bank, *Fifty Years after Bretton Woods: The Future of the IMF and the World Bank*. Washington, DC: IMF and World Bank.

Killick, Tony, Malik, Moazzam and Marcus, Manuel (1992a) 'What Can We Know About the Effects of IMF Programmes?', *World Economy* 15(5), September.

Killick, Tony, Malik, Moazzam and Marcus, Manuel (1992b) 'Country Experiences with IMF Programmes', *World Economy* 15(5), September.

Killick, Tony and Mwega, F.M. (1993) 'Monetary Policy in Kenya, 1967–88', in Sheila Page (ed.) *Monetary Policy in Developing Countries*. London and New York: Routledge.

Kimaro, S. (1988) 'Floating Exchange Rates in Africa', IMF *Working Paper* 88/47. Washington, DC: International Monetary Fund.

Lizondo, J. Saul and Montiel, Peter J. (1989) 'Contractionary Devaluation in Developing Countries: An Analytical Overview', *IMF Staff Papers* 36(1), March.

Loxley, John (1984) *The IMF and the Poorest Countries.* Ottawa: North–South Institute.

Loxley, John (1986) 'Alternative Approaches to Stabilization in Africa', in Gerald K. Helleiner (ed.) *Africa and the International Monetary Fund.* Washington, DC: International Monetary Fund.

Lynn, Robert and McCarthy, F. Desmond (1989) 'Recent Economic Performance of Developing Countries', *Working Paper* WPS 228. Washington, DC: World Bank, July.

Mansur, Ahsan H. (1989) 'Effects of a Budget Deficit on the Current Account Balance: The Case of the Philippines', in Mario I. Blejer and Ke-Young Chu (eds) *Fiscal Policy, Stabilization and Growth in Developing Countries.* Washington, DC: International Monetary Fund.

Martin, Matthew (1991) *The Crumbling Facade of African Debt Negotiations.* London and New York: Macmillan and St Martins.

Martin, Matthew and Mistry, Percy S. (1991) 'External Finance for African Development in the 1990s: Preliminary Project Report'. Oxford: Queen Elizabeth House, June (mimeo).

Matin, K.M. (1986) 'Bangladesh and the IMF: An Exploratory Study', *Research Monograph* No. 5. Dhaka: Bangladesh Institute of Development Studies, January.

Matin, K.M. (1990) 'Structural Adjustment Under the Extended Fund Facility: The Case of Bangladesh', *World Development* 18(5).

Mayer, Otto G. (1990) 'Policy Dialogue between the IMF, World Bank and Developing Countries', *Foreign Trade Review* XXV(2), July–September.

Mohammed, Azizali F. (1991) 'Recent Evolution of Fund Conditionality', in Jacob A. Frenkel and Morris Goldstein (eds) *International Financial Policy: Essays in Honor of J.J. Polak.* Washington, DC: International Monetary Fund.

Montes, Manuel F. (1987) *The Philippines*, Country Study No. 2, Stabilisation and Adjustment Programmes and Policies. Stockholm: WIDER.

Moore, Will H. and Scarritt, James R. (1990) 'IMF Conditionality and Polity Characteristics in Black Africa: An Exploratory Analysis', *Africa Today* 37(4).

Mosley, Paul, Harrigan, Jane and Toye, John (1991) *Aid and Power: The World Bank and Policy-Based Lending* (2 vols). London: Routledge.

Musalem, A. (1989) 'Private Investment in Mexico: An Empirical Analysis', World Bank *Working Paper* 183. Washington, DC: World Bank.

Nashashibi, K., Gupta, S., Liuksila, C., Lorie, H. and Mahler, W. (1992) *The Fiscal Dimensions of Adjustment in Low-Income Countries*, IMF Occasional Paper No. 95. Washington, DC: International Monetary Fund, April.

Ndulu, Benno (1987) *Tanzania*, Country Study 17, Stabilisation and Adjustment Programmes and Policies. Stockholm: WIDER.

Nelson, Joan M. (ed.) (1989) *Fragile Coalitions: The Politics of Economic Adjustment.* New Brunswick, N.J.: Overseas Development Council and Transaction Books.

Nelson, Joan M. (ed.) (1990) *Economic Crisis and Policy Choice: The Politics of Adjustment in the Third World*. Princeton, N.J.: Princeton University Press.

Özler, S. (1993) 'Have Commercial Banks Ignored History?', *American Economic Review* 83: 608–20.

Page, Sheila (ed.) (1993) *Monetary Policy in Developing Countries*. London and New York: Routledge.

Pastor, Manuel Jr (1987) 'The Effects of IMF Programs in the Third World: Debate and Evidence from Latin America', *World Development* 15(2), February.

Polak, Jacques J. (1957) 'Monetary Analysis of Income Formation and Payments Problems', *IMF Staff Papers* 5(4), November.

Polak, Jacques J. (1991) 'The Changing Nature of IMF Conditionality', *Princeton Essays in International Finance* No. 184. Princeton, N.J.: Princeton University, September.

Pradhan, S. and Swaroop, V. (1993) 'Public Spending and Adjustment', *Finance and Development* 30(3), September.

Quirk, Peter J., Christensen, B.V., Huh, K.-M. and Sasaki, T. (1987) *Floating Exchange Rates in Developing Countries*, IMF Occasional Paper No. 53. Washington, DC: International Monetary Fund, May.

Reichmann, Thomas M. (1978) 'The Fund's Conditional Assistance and the Problems of Adjustment: 1973–75', *Finance and Development* 15(12), December.

Reichmann, Thomas M. and Stillson, Richard D. (1978) 'Experience with Programs of Balance of Payments Adjustment: Stand-by Arrangements in the Higher Credit Tranches, 1963–72', *IMF Staff Papers* 25(2), June.

Reisen, Helmut and van Trotsenburg, Axel (1988) *Developing Country Debt: the Budgetary and Transfer Problem*. Paris: OECD.

Robinson, R.J. and Schmitz, L. (1989) 'Jamaica: Navigating through a Troubled Decade', *Finance and Development*, December.

Rodrik, Dani (1991) 'Policy Uncertainty and Private Investment in Developing Countries', *Journal of Development Economics* 36(2), October.

Ros, Jaime and Lustig, N. (1987) *Mexico*, Stabilisation and Adjustment Programmes and Policies. Stockholm: WIDER.

Rowlands, Dane (1994) 'The Response of New Lending to the IMF'. Ottawa: Carleton University (processed), April (mimeo).

Sachs, Jeffrey D. (1989a) 'Conditionality, Debt Relief, and the Developing Country Debt Crisis', in Jeffrey D. Sachs (ed.) *Developing Country Debt and Economic Performance, Vol. 1. International Financial System*, University of Chicago Press.

Sachs, Jeffrey D. (1989b) 'Strengthening IMF Programs in Highly Indebted Countries', in Catherine Gwin and Richard Feinberg (eds) *The International Monetary Fund in a Multipolar World: Pulling Together*, US–Third World Policy Perspectives No. 13. Washington, DC: Overseas Development Council.

Sachs, Jeffrey D. (1989c) 'New Approaches to the Latin American Debt Crisis', *Princeton Studies in International Finance* No. 174. Princeton, N.J.: Princeton University, July.

Schadler, Susan, Rozwadowski, Franek, Tiwari, Siddharth and Robinson, David O. (1993) *Economic Adjustment in Low-Income Countries: Experiences*

*Under the Enhanced Structural Adjustment Facility*, Occasional Paper 106. Washington, DC: International Monetary Fund, September.

Schneider, F. and Frey, B.S. (1985) 'Economic and Political Determinants of Foreign Direct Investment', *World Development* 13(2), February.

Serven, Luis and Solimano, Andres (eds) (1993) *Striving for Growth After Adjustment: The Role of Capital Formation*. Washington, DC: World Bank.

Sharpley, J. (1984) 'Jamaica, 1972–80', in Tony Killick (ed) *The IMF and Stabilisation: Developing Country Experiences*. London: Gower Publishing Company and Overseas Development Institute.

Sidell, Scott R. (1988) *The IMF and Third World Instability: Is There a Connection?* Basingstoke: Macmillan.

Spraos, John (1986) 'IMF Conditionality: Ineffectual, Inefficient, Mistargeted', *Princeton Essays in International Finance* No. 166. Princeton, N.J.: Princeton University Press, December.

Stiglitz, Joseph E. (1992) 'Capital Markets and Economic Fluctuations in Capitalist Economies', *European Economic Review* 36(2/3).

Stiles, Kendall W. (1991), *Negotiating Debt: The IMF Lending Process*. Boulder, Colo.: Westview Press.

Stuart, Brian C. (1991) 'Country Experiences and Programming Issues – An Overview'. Washington, DC: International Monetary Fund (mimeo).

Tanzi, Vito (1989) 'Fiscal Policy, Growth and the Design of Stabilization Programs', in Mario I. Blejer and Ke-Young Chu (eds) *Fiscal Policy, Stabilization and Growth in Developing Countries*. Washington, DC: International Monetary Fund.

Taylor, Lance (1988) *Varieties of Stabilization Experience*. Oxford: Clarendon Press.

Taylor, Lance (ed.) (1993) *The Rocky Road to Reform*. Cambridge, Mass.: MIT Press.

Thomas, Vinod, Chhibber, Ajay, Dailami, Mansoor and de Melo, Jaime (eds) (1991) *Restructuring Economies in Distress: Policy Reform and the World Bank*. Oxford: Oxford University Press.

UNICEF Manila (1988) 'Redirecting Adjustment Programmes Towards Growth and the Protection of the Poor: The Philippine Case', in G.A. Cornia *et al.* (eds) *Adjustment with a Human Face: Ten Country Case Studies*. Oxford: Clarendon Press and UNICEF.

Wade, Robert (1990) *Governing the Market: Economic Theory and the Role of Government in East Asian Industrialization*. Princeton, N.J.: Princeton University Press.

Walters, Sir Alan (1994) 'Do We Need the IMF and the World Bank?', *Current Controversies* No. 10. London: Institute of Economic Affairs.

Williamson, John (1980) 'Economic Theory and IMF Policies', in K. Brunner and A.H. Meltzer (eds) *Monetary Institutions and the Policy Process*, Carnegie–Rochester Conference Series on Public Policy. Washington, DC: North-Holland.

World Bank (1988) *World Development Report, 1988*. New York: Oxford University Press.

World Bank (1989) *Adjustment Lending: An Evaluation of Ten Years of Experience*. Washington, DC: World Bank.

World Bank (1992) *World Bank Structural and Sectoral Adjustment Operations: The Second OED Overview*, Operations Evaluation Department Report No. 10870. Washington, DC: World Bank, June.

World Bank (1993) *The East Asian Miracle: Economic Growth and Public Policy.* New York: Oxford University Press.

World Bank (1994a) *Global Economic Prospects, 1994*. Washington, DC: World Bank.

World Bank (1994b) *Adjustment in Africa: Reforms, Results and the Road Ahead.* New York: Oxford University Press.

Zulu, Justin B. and Nsouli, Saleh M. (1985) *Adjustment Programs in Africa: The Recent Experience*, IMF Occasional Paper No. 34. Washington, DC: International Monetary Fund.

# INDEX

# DO IMF POLICIES HELP DEVELOPING COUNTRIES?

As linchpin of the global financial system, the International Monetary Fund provides balance of payments support, chiefly to developing countries, conditional on strict remedial policy measures.Its approach to policy remains highly controversial, however. While the Fund claims it has adapted, critics allege its policies are harshly doctrinaire, imposing hardships on already poor people. For the critics, the half-century of its existence is 'fifty years too long' and radical change is essential.

This book examines the arguments, tracing the extent of Fund adaptation, presenting major new evidence on the consequences of Fund programmes, and considering its future role.

**Tony Killick** is Senior Research Fellow with the Overseas Development Institute, London, and visiting Professor of the University of Surrey.

Development Policy Studies

Edited by John Farrington and Tony Killick for the Overseas Development Institute.

**Development/Economics/Geography**

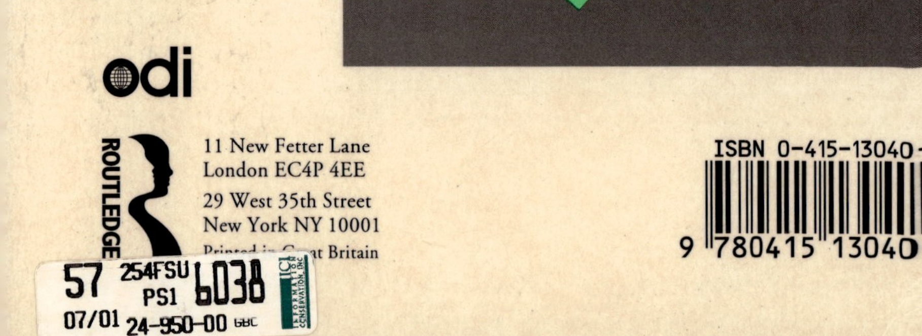


11 New Fetter Lane
London EC4P 4EE

29 West 35th Street
New York NY 10001

Printed in Great Britain

ISBN 0-415-13040-9

9 780415 130400

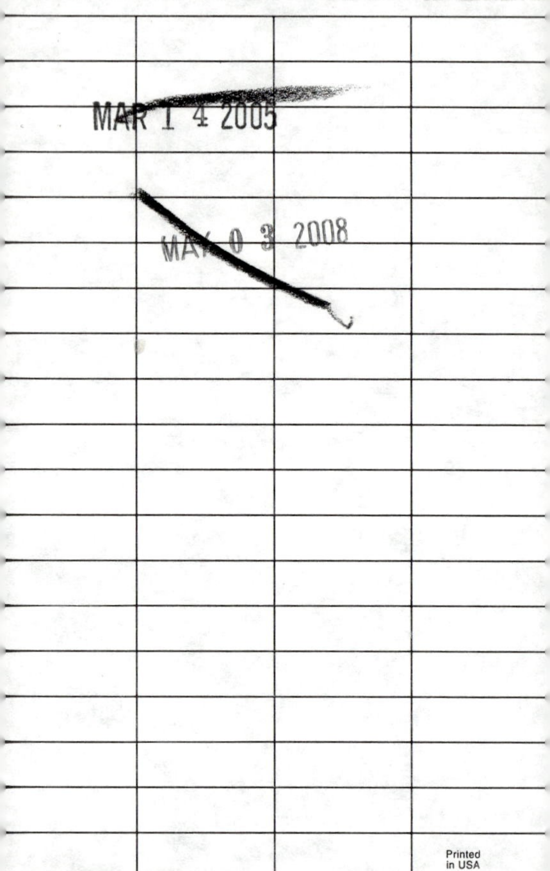
MAR 1 4 2005
MAY 0 3 2008